CONGRESSIONAL POLICIES, PRACTICES AND PROCEDURES

WATER QUALITY

BACKGROUND AND ISSUES FOR CONGRESS

Congressional Policies, Practices and Procedures

Additional books in this series can be found on Nova's website under the Series tab.

Water Resource Planning, Development and Management

Additional books in this series can be found on Nova's website under the Series tab.

CONGRESSIONAL POLICIES, PRACTICES AND PROCEDURES

WATER QUALITY

BACKGROUND AND ISSUES FOR CONGRESS

EDWARD SAUNDERS
AND
LEONA P. MORGAN
EDITORS

snova
New York

Library of Congress Cataloging-in-Publication Data

ISBN: 978-1-53614-809-1

Published by Nova Science Publishers, Inc. † New York

CONTENTS

PREFACE

Much progress has been made in achieving the ambitious goals that Congress established in 1972 in the Clean Water Act (CWA) to restore and maintain the chemical, physical, and biological integrity of the nation's waters. However, long-standing problems persist, and new problems have emerged. Water quality problems are diverse, ranging from pollution runoff from farms and ranches, city streets, and other diffuse or "nonpoint" sources, to toxic substances discharged from factories and sewage treatment plants. Since the early 2000s, increased oil and gas production across the nation has resulted in a corresponding increase in wastewater that must be managed, reused, or disposed of properly. In particular, the hydraulic fracturing process has also raised concerns about potential effects to human health and the environment, including the potential contamination of underground drinking water sources by injecting wastewater associated with the production of oil and gas.

Chapter 1 - Much progress has been made in achieving the ambitious goals that Congress established in 1972 in the Clean Water Act (CWA) to restore and maintain the chemical, physical, and biological integrity of the nation's waters. However, long-standing problems persist, and new problems have emerged. Water quality problems are diverse, ranging from pollution runoff from farms and ranches, city streets, and other diffuse or "nonpoint" sources, to toxic substances discharged from factories and sewage treatment plants.

There is little agreement among stakeholders about what solutions are needed, whether legislation is required to address the nation's remaining water pollution problems, or whether regulatory authorities should be reduced. For some time, efforts to comprehensively amend the CWA have stalled as interests have debated whether and exactly how to change the law. Congress has instead focused legislative attention on enacting narrow bills to extend or modify selected CWA programs, but not comprehensive proposals.

Programs that regulate activities in wetlands have been of particular interest recently, especially CWA Section 404, which has been criticized by landowners for intruding on private land-use decisions and imposing excessive economic burdens. Environmentalists view this regulatory program as essential for maintaining the health of wetland ecosystems, and they are concerned about court rulings that have narrowed regulatory protection of wetlands. Many stakeholders desire clarification of the act's regulatory jurisdiction, but they differ on what solutions are appropriate. On May 27, 2015, the Environmental Protection Agency (EPA) and the Army Corps of Engineers finalized a rule intended to clarify jurisdictional issues, but interpretive questions about the rule remain controversial inside and outside of Congress.

Another prominent water quality issue for some time has concerned financial aid for municipal wastewater treatment projects. House and Senate committees have approved bills to reauthorize CWA assistance on several occasions since the 107th Congress, but, for various reasons, no legislation other than appropriations was enacted. At issue has been the role of the federal government in assisting states and cities in meeting needs to rebuild, repair, and upgrade wastewater treatment systems, especially in light of capital costs that are projected to be nearly $300 billion over the next 20 years. The 113th Congress agreed to legislation that creates a pilot program to provide federal loans for water infrastructure projects (P.L. 113-121), but the program has not yet received funds to begin making loans. The same legislation also revised certain of the water infrastructure financing provisions of the CWA.

A number of other water quality issues have been the subject of congressional oversight and legislation, with some legislators highly critical of EPA's recent regulatory initiatives and others more supportive. In several cases, policymakers have sought to curtail water quality protection initiatives under the CWA following court rulings that expanded the regulatory scope of the law. Among the topics of particular interest has been regulation of surface coal mining activities in Appalachia. Congressional interest in this and other topics has been reflected in specific legislative proposals and debate over policy provisions of bills providing appropriations for EPA. Members from both parties have raised questions about the cost-effectiveness of some of EPA's actions and whether the agency has exceeded its authority. In the 114th Congress, scrutiny of EPA initiatives has continued to be intense.

Chapter 2 - Section 401 of the Clean Water Act (CWA) requires that an applicant for a federal license or permit provide a certification that any discharges from the facility will comply with the act, including state-established water quality standard requirements. Disputes have arisen over the states' exercise of this authority in protecting water quality. For the most part, the debate over the Section 401 certification issue has been between states and hydropower interests. A 1994 Supreme Court decision, which upheld the states' authority in this area, dismayed development and hydropower interest groups. The Court revisited these issues in a 2006 ruling that unanimously upheld the authority of states to condition hydropower licenses by exercising Section 401. The dispute between states and industry groups about Section 401 authority has been a legislative issue on several occasions, but Congress has not modified the provision's scope. In the 114th Congress, Senate and House committees are considering a number of energy policy proposals, including legislation that some states believe could impinge on states' authority under CWA Section 401 concerning hydropower projects.

In addition, there has been interest in clarifying whether Section 401 certification applies to nonpoint source discharges, such as rainfall runoff, as well as point source discharges from pipes or ditches. This question was raised in lawsuits in Oregon, where a federal court ruled in 1998 and again in 2008 that Section 401 does not apply to nonpoint source discharges. Still, some interests continue to favor a broad reading of 401 that would apply to both nonpoint and point sources of pollutant discharges.

Chapter 3 - The scope of waters that are properly the subject of federal water pollution legislation has been the subject of long-standing consideration by all three branches of the federal government, particularly in the aftermath of the 1972 amendments to the Federal Water Pollution Control Act, commonly referred to as the Clean Water Act. In a shift from early water pollution legislation, those amendments eliminated the requirement that the federally regulated waters—known as jurisdictional waters—must be navigable in the traditional sense, meaning that they are capable of being used by vessels in interstate commerce. Rather than use classical tests of navigability, the amendments redefined "navigable waters" for purposes of the Clean Water Act's jurisdiction to include "the waters of the United States, including the territorial seas." Disputes over the proper meaning of that phrase have been ongoing.

Some courts and commentators also disagree on how the scope of federal jurisdictional waters changed over time as a result of interpretative approaches taken by the federal agencies responsible for administering the Clean Water Act—the Environmental Protection Agency (EPA) and the U.S. Army Corps of Engineers (Corps). While some believe EPA and the Corps consistently expanded the meaning of "waters of the United States," others contend that, in recent years, the agencies have construed the term in a narrower fashion than permitted under the Clean Water Act. In 2015, the Corps and EPA issued a new rule, known as the Clean Water Rule that substantially redefined "waters of the United States" in the agencies' regulations for the first time in more than two decades. Some observers disagree on whether the Clean Water Rule constitutes an expansion of jurisdiction over waters not previously regulated. This report provides context for this debate by examining the history of major changes to the meaning of "waters of the United States" as expressed in federal regulations, legislation, agency guidance, and case law.

The Clean Water Act uses the phrase "waters of the United States," but it does not include a statutory definition of that term. The long-standing disagreement over the meaning of that phrase has centered on the degree to which the Clean Water Act should be interpreted as covering the widest amount of "waters" that could permissibly be federally regulated under the Constitution, or whether that term should be interpreted in a more limited fashion.

Federal authority to regulate waters within the United States primarily derives from the Commerce Clause, and accordingly, federal laws and regulations concerning waters of the United States cannot cover matters which exceed that constitutional source of authority. During the first two decades after the passage of the Clean Water Act, courts generally interpreted the act as having a wide jurisdictional reach. In recent decades, however, the Supreme Court has emphasized that "the grant of authority to Congress under the Commerce Clause, though broad, is not unlimited." This modern Commerce Clause jurisprudence has informed federal courts' approach to interpreting which "waters" are subject to the Clean Water Act.

Most recently, courts have taken up legal challenges to the Clean Water Rule. On October 9, 2015, the United States Court of Appeals for the Sixth Circuit stayed its enforcement, and the House version of the FY2017 Interior-Environment appropriations bill (H.R. 5538) would block its application by prohibiting the use of appropriated funds to implement changes to the meaning of jurisdictional waters beyond those that were in effect on October 1, 2012.

Chapter 4 - In *National Association of Manufacturers (NAM) v. Department of Defense*, the Supreme Court held that legal challenges to the 2015 Clean Water Rule issued by the Environmental Protection Agency (EPA) and Army Corps of Engineers (Corps) must be litigated in multiple federal district courts rather than in a consolidated case in the United States Court of Appeals for the Sixth Circuit (Sixth Circuit). Since October 2015, a Sixth Circuit order had stayed implementation of the Clean Water Rule on a nationwide basis pending the court's determination of its jurisdiction over the challenge to the Rule. But the Supreme Court's *NAM* decision, which requires the Sixth Circuit to dismiss its case, will result in the elimination of that order, and potentially could complicate legal challenges to the Clean Water Rule and the Trump Administration's effort to rescind and replace it.

Chapter 5 - The principal federal program to aid municipal wastewater treatment plant construction is authorized in the Clean Water Act (CWA). Established as a grant program in 1972, it now capitalizes state loan programs through the clean water state revolving loan fund (CWSRF) program. Since FY1972, appropriations have totaled $97 billion.

In 1996, Congress amended the Safe Drinking Water Act (SDWA, P.L. 104-182) to authorize a drinking water SRF program to help systems finance projects needed to comply with drinking water regulations and to protect public health. Since FY1997, appropriations for the drinking water state revolving loan fund (DWSRF) program have totaled $21 billion.

The U.S. Environmental Protection Agency (EPA) administers both SRF programs, which annually distribute funds to the states for implementation. Funding amounts are specified in the State and Tribal Assistance Grants (STAG) account of EPA annual appropriations acts. The combined appropriations for wastewater and drinking water infrastructure assistance have represented 25%-32% of total funds appropriated to EPA in recent years.

Prior to CWA amendments in 1987 (P.L. 100-4), Congress provided wastewater grant funding directly to municipalities. The federal share of project costs was generally 55%; state and local governments were responsible for the remaining 45%. The 1987 amendments replaced the traditional grant program with the SRF program. Local communities are now often responsible for 100% of project costs, rather than 45%, because they are required to repay loans to states. The greater financial burden of the act's loan program on some cities has caused some to seek continued grant funding.

This concern has been particularly evident in the appropriations process where, for a number of years, Congress reserved as much as 30% of funds in the STAG account for special purpose grants directed to specified communities. Most of the funded projects were not authorized in the CWA or SDWA. Many state water quality officials and state infrastructure financing officials objected to this practice, as did EPA, since it reduced the amount of funding for state SRF programs. Due to general opposition to congressional earmarking, this practice has not occurred since FY2011.

Although the CWSRF and DWSRF have largely functioned as loan programs, both allow the implementing state agency to provide "additional subsidization" under certain conditions. Since its amendments in 1996, the SDWA has *authorized* states to use up to 30% of their DWSRF capitalization grants to provide additional assistance, such as forgiveness of loan principal or negative interest rate loans, to help disadvantaged communities (as determined by the state). Congress amended the CWA in 2014, adding similar provisions to the CWSRF program. In addition, appropriations acts in recent years have *required* states to use minimum percentages of their allotted SRF grants to provide additional subsidization.

The final full-year appropriations for FY2018 were enacted as part of the Consolidated Appropriations Act, 2018, signed by President Trump on March 23, 2018 (P.L. 115-141). Compared to FY2017, funding for EPA water infrastructure programs increased by 24% in FY2018. The act provided $1.694 billion for the CWSRF and $1,163 million for the DWSRF program, increases of $300 million to both programs compared to FY2017. The act also provided $63 million for the WIFIA program, more than doubling the FY2017 appropriation.

The Trump Administration's FY2019 budget request proposes the same amounts for the CWSRF and DWSRF programs as the FY2017 appropriation. However, the request proposes a 45% reduction in the state categorical grants compared to FY2018 levels.

Chapter 6 - Since the early 2000s, increased oil and gas production has resulted in an increase in wastewater that must be managed properly. The majority of wastewater from oil and gas production is injected into underground wells known as class II wells. These wells are regulated to protect drinking water sources under EPA's UIC class II well program and approved state class II programs. EPA oversees state programs, and EPA regions manage programs in states without approval.

GAO was asked to review EPA's oversight of programs' inspection and enforcement information and activities. This report examines the extent to which EPA has collected inspection and enforcement information and conducted oversight activities needed to assess that class II programs protect underground sources of drinking water. GAO reviewed federal and state laws and regulations and EPA guidance and analyzed a nongeneralizable sample of significant violations. GAO interviewed EPA and state officials from programs in a nongeneralizable sample of eight states selected based on shale oil and gas regions, among other factors.

Chapter 7 - In April 2014, the city of Flint, Michigan, stopped purchasing treated water from the city of Detroit and began using the Flint River as its water source without providing corrosion control treatment. (A key source of lead in drinking water is the corrosion of plumbing materials in the distribution system and households.) While some water quality changes were quickly apparent, elevated lead concentrations were identified over a longer period through monitoring conducted by the city and others and detections of elevated blood lead levels in children. On October 1, 2015, city officials urged residents to stop drinking the water. On October 16, Flint reconnected to Detroit's water and advised residents not to use unfiltered tap water. The city and the governor of Michigan each declared a state of emergency. President Obama issued an emergency declaration on January 16, 2016.

On January 21, 2016, the Environmental Protection Agency (EPA) issued an emergency order directing the city and state to take immediate actions to address concerns over the city's water system. Requirements include, among others, that the city and state re-optimize corrosion control, post online lead monitoring results and weekly reports, and ensure the city's capacity to operate the system in compliance with federal regulations. EPA's current Flint responses include providing technical assistance for water testing and treatment, conducting water monitoring, and identifying lead service line locations.

Chapter 8 - Scientific research indicates that in recent years, the frequency and geographic distribution of harmful algal blooms (HABs) have been increasing nationally and globally. Because the impacts of

HABs can be severe and widespread—often with interstate implications—these issues have been a perennial interest for Congress. While algal communities are natural components of healthy aquatic ecosystems, under certain conditions (e.g., increased temperatures and nutrient concentrations), algae may grow excessively, or "bloom," and produce toxins that can harm human health, animals, aquatic ecosystems, and the economy.

In 2014, a cyanobacterial HAB in Lake Erie affected the drinking water for more than 500,000 people in Toledo, Ohio. In 2016, a massive HAB in Florida's Lake Okeechobee negatively impacted tourism and aquatic life. HABs have been recorded in every state and have become a concern nationwide.

Many types of algae can cause HABs in freshwater systems. The most frequent and severe blooms involve the proliferation of cyanobacteria. Some cyanobacteria species can produce toxins—cyanotoxins—that can cause mild to severe health effects in humans and kill aquatic life and other animals.

HABs can also contribute to deteriorating water quality and ecosystem health. As masses of cyanobacteria or other algae die and decompose, they consume oxygen, sometimes forming "dead zones" where life cannot survive. These areas can kill fish and organisms, such as crabs and clams, and have detrimental economic effects.

Scientists widely consider nutrient enrichment to be a key cause of HAB formation. While nutrients are essential to plants and natural parts of aquatic ecosystems, excessive amounts can overstimulate algal growth. Sources include point sources (e.g., municipal wastewater discharges) and nonpoint sources (e.g., fertilizer runoff from agricultural and urban areas).

Congress, federal agencies, and states have taken steps to address HABs and nutrients that contribute to their occurrence. The Harmful Algal Bloom and Hypoxia Research and Control Act of 1998 (HABHRCA), as amended, established an interagency task force, required the task force to prepare reports and plans addressing marine and freshwater HABs, and authorized funding for research, education, monitoring activities, etc.

In December 2016, the Environmental Protection Agency (EPA) used its authority under the Clean Water Act (CWA) to propose water quality criteria for two algal toxins in waters used for recreational purposes. States use such criteria when developing water quality standards— measures that describe the desired condition or level of protection of a water body and what is needed for protection.

Further, EPA has emphasized the need to reduce nutrient pollution from *all* sources to reduce public health and environmental impacts associated with HABs. The CWA does not authorize EPA to regulate all sources. It authorizes EPA to regulate point (direct) sources of nutrients but does not authorize EPA to regulate nonpoint (diffuse) sources of nutrient pollution.

Some states have developed guidelines for algal toxins, primarily for use in guiding swimming advisories. Also, states have listed waters as impaired, or not meeting water quality standards, for algal blooms or algal toxins. Some of these states have begun to develop Total Maximum Daily Loads (TMDLs)—essentially pollution budgets—to address them. Most states have identified nutrient-related pollution as a priority to be addressed by their TMDLs and/or alternative restoration plans. States rely heavily on financial assistance from EPA in implementing these plans and more broadly, in addressing nonpoint source pollution that leads to degraded water quality and HAB formation. Congress has long provided financial assistance through EPA for regional, state, and local programs through CWA Sections 106 and 319 planning grants, geographic programs (such as the Chesapeake Bay and Great Lakes), and other sources. The President's FY2018 budget request for most of these programs is either eliminated or significantly reduced.

Congress continues to show interest in addressing HABs. This interest has largely focused on funding research to close research gaps identified by scientists and decisionmakers and to coordinate the efforts of federal agencies and their partners to study and address HABs.

In: Water Quality
Editors: Edward Saunders et al.
ISBN: 978-1-53614-809-1

Chapter 1

WATER QUALITY ISSUES IN THE 114TH CONGRESS: AN OVERVIEW*

Claudia Copeland

SUMMARY

Much progress has been made in achieving the ambitious goals that Congress established in 1972 in the Clean Water Act (CWA) to restore and maintain the chemical, physical, and biological integrity of the nation's waters. However, long-standing problems persist, and new problems have emerged. Water quality problems are diverse, ranging from pollution runoff from farms and ranches, city streets, and other diffuse or "nonpoint" sources, to toxic substances discharged from factories and sewage treatment plants.

There is little agreement among stakeholders about what solutions are needed, whether legislation is required to address the nation's remaining water pollution problems, or whether regulatory authorities should be reduced. For some time, efforts to comprehensively amend the CWA have stalled as interests have debated whether and exactly how to change the law. Congress has instead focused legislative attention on

---

* This is an edited, reformatted and augmented version of Congressional Research Service, Publication R43867, dated January 5, 2016.

enacting narrow bills to extend or modify selected CWA programs, but not comprehensive proposals.

Programs that regulate activities in wetlands have been of particular interest recently, especially CWA Section 404, which has been criticized by landowners for intruding on private land-use decisions and imposing excessive economic burdens. Environmentalists view this regulatory program as essential for maintaining the health of wetland ecosystems, and they are concerned about court rulings that have narrowed regulatory protection of wetlands. Many stakeholders desire clarification of the act's regulatory jurisdiction, but they differ on what solutions are appropriate. On May 27, 2015, the Environmental Protection Agency (EPA) and the Army Corps of Engineers finalized a rule intended to clarify jurisdictional issues, but interpretive questions about the rule remain controversial inside and outside of Congress.

Another prominent water quality issue for some time has concerned financial aid for municipal wastewater treatment projects. House and Senate committees have approved bills to reauthorize CWA assistance on several occasions since the 107th Congress, but, for various reasons, no legislation other than appropriations was enacted. At issue has been the role of the federal government in assisting states and cities in meeting needs to rebuild, repair, and upgrade wastewater treatment systems, especially in light of capital costs that are projected to be nearly $300 billion over the next 20 years. The 113th Congress agreed to legislation that creates a pilot program to provide federal loans for water infrastructure projects (P.L. 113-121), but the program has not yet received funds to begin making loans. The same legislation also revised certain of the water infrastructure financing provisions of the CWA.

A number of other water quality issues have been the subject of congressional oversight and legislation, with some legislators highly critical of EPA's recent regulatory initiatives and others more supportive. In several cases, policymakers have sought to curtail water quality protection initiatives under the CWA following court rulings that expanded the regulatory scope of the law. Among the topics of particular interest has been regulation of surface coal mining activities in Appalachia. Congressional interest in this and other topics has been reflected in specific legislative proposals and debate over policy provisions of bills providing appropriations for EPA. Members from both parties have raised questions about the cost-effectiveness of some of EPA's actions and whether the agency has exceeded its authority. In the 114th Congress, scrutiny of EPA initiatives has continued to be intense.

INTRODUCTION

Much progress has been made in achieving the ambitious goals that Congress established more than 40 years ago to restore and maintain the chemical, physical, and biological integrity of the nation's waters. However, long-standing problems persist, and new problems have emerged. Water quality problems are diverse, ranging from pollution runoff from farms and ranches, city streets, and other diffuse or "nonpoint" sources, to "point" source discharges of metals and organic and inorganic toxic substances from factories and sewage treatment plants.

The principal law that deals with polluting activity in the nation's streams, lakes, estuaries, and coastal waters is the Federal Water Pollution Control Act (P.L. 92-500, enacted in 1972), commonly known as the Clean Water Act, or CWA. It consists of two major parts: regulatory provisions that impose progressively more stringent requirements on industries and cities to abate pollution and meet the statutory goal of zero discharge of pollutants; and provisions that authorize federal financial assistance for municipal wastewater treatment plant construction. Both parts are supported by research activities, plus permit and enforcement provisions. Programs at the federal level are administered by the Environmental Protection Agency (EPA); state governments have primary day-to-day responsibilities to implement CWA programs through standard-setting, permitting, enforcement, and administering financial assistance programs. Local governments also have important roles in implementing water quality protection programs, such as building and operating municipal wastewater treatment plants and regulating local pollution sources.[1]

The water quality restoration objective declared in the 1972 act was accompanied by statutory goals to attain, wherever possible, waters deemed "fishable and swimmable" by 1983 and to eliminate the discharge of pollutants into navigable waters by 1985. Although those goals have not been fully achieved, considerable progress has been made, especially in controlling conventional pollutants (suspended solids, bacteria, and

[1] For further information, see CRS Report RL30030, *Clean Water Act: A Summary of the Law*, by Claudia Copeland.

oxygen-consuming materials) discharged by industries and sewage treatment plants.

Progress has been mixed in controlling discharges of toxic pollutants (heavy metals, inorganic and organic chemicals), which are more numerous and can harm human health and the environment even when present in very small amounts—at the parts-per-billion level. Moreover, efforts to control pollution from diffuse sources, termed nonpoint source pollution (rainfall runoff and snowmelt from urban, suburban, and agricultural areas, for example), are more recent, given the earlier emphasis on "point source" pollution (discharges from industrial facilities and municipal wastewater treatment plants). Overall, data reported by EPA and states indicate that 44% of river and stream miles assessed by states and 64% of assessed lake acres do not meet applicable water quality standards and are impaired for one or more desired uses.[2] In 2006 EPA issued an assessment of streams and small rivers, reporting that 67% of U.S. stream miles are in poor or fair condition and that nutrients and streambed sediments have the largest adverse impact on the aquatic species in these waters.[3] A similar assessment of the health of the nation's lakes, issued in 2009, found that 56% are in good biological condition, but that about 20% of lakes have high levels of phosphorus or nitrogen and are more likely to have poor biological health than lakes without excess nutrients.[4] Approximately 95,000 lakes and 544,000 river miles in the United States are under fish-consumption advisories (including 100% of the Great Lakes and their connecting waters), due to chemical contaminants, and one-third of shellfishing beds are closed or restricted due to toxic pollutant contamination. Mercury is a contaminant of growing concern—in 2010, approximately 16.3 million lake acres and 1.14 million river miles were

---

[2] U.S. Environmental Protection Agency, *National Water Quality Inventory: Report to Congress, 2004 Reporting Cycle*, EPA 841-R-08-001, January 2009, http://water.epa.gov/lawsregs/guidance/cwa/305b/2004report_index.cfm.

[3] U.S. Environmental Protection Agency, *Wadeable Streams Assessment: A Collaborative Survey of the Nation's Streams*, EPA 841-B-06-002, December 2006, http://www.epa.gov/owow/streamsurvey/.

[4] U.S. Environmental Protection Agency, Office of Water and Office of Research and Development, *National Lakes Assessment: A Collaborative Survey of the Nation's Lakes*, EPA-R-09-001, 2009.

under fish or shellfish consumption advisory because of elevated mercury levels. Mercury concentrations in game fish exceed health-based limits in about half of U.S. lakes.[5]

The last major amendments to the CWA were the Water Quality Act of 1987 (P.L. 100-4). That legislation culminated six years of congressional efforts to extend and revise the act and was the most comprehensive amendments since 1972. Authorizations of appropriations for some programs provided in P.L. 100-4, such as grant assistance to states, research, and general EPA support, expired in FY1990 and FY1991. Authorizations for wastewater treatment project funding expired in FY1994. None of these programs has lapsed, however, as Congress has continued to appropriate funds to implement them. EPA, states, industry, and other citizens continue to implement the 1987 legislation.

The Clean Water Act has been viewed as one of the most successful environmental laws in terms of achieving its statutory goals, which have been widely supported by the public. Lately, however, some have questioned whether additional actions to achieve further benefits are worth the costs, especially in view of the continuing problems of the U.S. economy. Criticism has come from industry, which has been the long-standing focus of the act's regulatory programs and often opposes imposition of new stringent and costly requirements. Criticism also has come from developers and property rights groups who contend that federal regulations (particularly the act's wetlands permit program) are a costly intrusion on private land-use decisions. States and cities have traditionally supported water quality programs and federal funding to assist them in carrying out the law, but many have opposed CWA measures that they fear might impose new unfunded mandates. Many environmental groups believe that strengthening of the law is needed to maintain progress achieved to date and to address remaining water quality problems.

---

[5] Ibid.

LEGISLATIVE AND OVERSIGHT ISSUES

October 2012 marked the 40th anniversary of passage of the Clean Water Act and 25 years since the last major amendments to the law were enacted. While there has been measurable clean water progress as a result of the act, observers and analysts agree that significant water pollution problems remain. However, there is less agreement about what solutions are needed and whether new legislation is required. Several key water quality issues exist: what additional actions, if any, should be taken to implement existing provisions of the law; whether additional steps are necessary to achieve overall goals of the act that have not yet been attained; how to ensure that progress made to date is not lost through diminished attention to water quality needs; whether existing regulatory authorities should be reduced; and what is the appropriate federal role in guiding and paying for clean water infrastructure and other activities. For some time, efforts to comprehensively amend the act have stalled as interests have debated whether and exactly how to change the law. Many issues that might be addressed involve making difficult tradeoffs between impacts on different sectors of the economy; taking action when there is technical or scientific uncertainty; and allocating governmental responsibilities among federal, state, local, and tribal entities for implementing the law.

These factors partly explain why Congress has recently focused legislative attention on narrow bills to extend or modify selected CWA programs, rather than taking up comprehensive proposals. Other factors also have been at work. These include a lack of legislative initiatives by the Administration on clean water issues (neither the Clinton nor the Bush Administration proposed CWA legislation, nor has the Obama Administration); and the high economic cost of addressing water infrastructure issues.

After the 2010 election, congressional attention turned significantly to oversight and legislation focused on criticism of EPA regulatory activities—particularly in the House, which passed a number of bills to limit EPA's regulatory authority. The Senate did not act on these measures.

The 112th Congress enacted two bills that amend the CWA. One extended the moratorium for CWA permitting of certain vessels for an additional year, until December 18, 2014 (P.L. 112-213), and the other extended authorization of funds for the Lake Pontchartrain Basin program in Section 121 of the act through FY2017 (P.L. 112-237). The 113th Congress enacted several bills with CWA provisions:

- Provisions of water resource legislation with some amendments to CWA Title VI, plus a pilot program for water infrastructure financing (see "WIFIA Pilot Program and SRF Amendments in P.L. 113-121" below);
- As part of the 2014 farm bill, legislation exempting most silviculture activity from requiring a CWA permit;[6]
- Legislation providing an additional three-year moratorium for CWA permitting of certain vessels;[7] and
- A bill eliminating a number of statutorily required reports to Congress, including one CWA report (P.L. 113-188).

Two CWA issues that have been the focus of much of legislators' interest in recent Congresses— regulatory protection of wetlands and water infrastructure financing—are discussed next.

Regulatory Protection of Wetlands

How best to protect the nation's remaining wetlands and regulate activities taking place in wetlands has become one of the most contentious environmental policy issues. Much of the debate has focused on the CWA, which contains a key wetlands regulatory tool. The permit program in CWA Section 404 requires landowners or developers to obtain permits for

---

[6] For discussion, see CRS Report R42883, *Water Quality Issues in the 113th Congress: An Overview*, by Claudia Copeland.

[7] See CRS Report R42142, EPA's Vessel General Permits: Background and Issues, by Claudia Copeland.

disposal of dredged or fill material that is generated by construction or similar activity into navigable waters of the United States, including wetlands. Section 404 has evolved through judicial interpretation and regulatory change to become one of the principal federal tools used to protect wetlands, although that term appears only once in Section 404 itself and is not defined there. At the same time, its implementation has come to be seen as intrusive and burdensome to those whose activities it regulates. At issue today is how to address criticism of the Section 404 regulatory program while achieving goals of wetlands protection in the context of meeting the objectives of the CWA.[8] Recently, the issue of wetlands management and protection has been central to controversy surrounding an Obama Administration regulatory proposal to define "waters of the United States," that is, which surface waters and wetlands are subject to the CWA's regulatory requirements and protection (see "2015 Revised Rule" below).

Unlike the rest of the act, the permit aspects of Section 404 are administered by the U.S. Army Corps of Engineers, rather than EPA, although the Corps uses environmental guidance jointly developed with EPA to evaluate permit applications. Also, the act authorizes EPA to veto a 404 permit that does not meet the law's requirements. Other federal agencies, including the Fish and Wildlife Service (FWS) and Natural Resource Conservation Service (NRCS), have more limited roles in the Corps' permitting decisions. Tension has existed for many years between the regulation of activities in wetlands under Section 404 and related laws, on the one hand, and the desire of landowners to develop property that may include wetlands, on the other hand. The conflicts over wetlands regulation have for the most part been addressed in administrative and judicial proceedings, as Congress has not amended Section 404 since 1977, when it provided exemptions for categories of routine activities, such as normal farming and forestry. Controversy has grown over the extent of federal jurisdiction and impacts on private property, burdens and delay of permit procedures, and roles of federal agencies and states in issuing permits.

---

[8] For additional information, see CRS Report RL33483, *Wetlands: An Overview of Issues*, by Claudia Copeland.

Judicial Proceedings Involving Section 404

One issue involving long-standing controversy and litigation is whether isolated waters are properly within the jurisdiction of Section 404. Waters and wetlands that appear to be isolated— e.g., they are not physically adjacent to navigable surface waters—or streams that are wet only for portions of the year may appear to provide only some of the values for which wetlands are protected, such as flood control or water purification, even if they meet the technical definition of a wetland.[9] Questions about whether such waters and wetlands are jurisdictional for CWA purposes have been extensively litigated.

SWANCC and Rapanos

In 2001, the Supreme Court ruled on the question of whether the CWA provides the Corps and EPA with authority over isolated waters. The Court's 5-4 ruling in *Solid Waste Agency of Northern Cook County (SWANCC) v. U.S. Army Corps of Engineers*[10] held that the Corps' denial of a 404 permit for a disposal site on isolated wetlands solely on the basis that migratory birds use the site exceeded the authority provided in the act.

In 2006, the Supreme Court revisited issues related to the extent of CWA jurisdiction in two consolidated cases brought by landowners (*Rapanos v. United States*; and *Carabell v. U.S. Army Corps of Engineers*) seeking to narrow the scope of the 404 permit program as it applies to development of wetlands. The issue in both cases had to do with the reach of the CWA to cover "waters" that were not navigable waters in the traditional sense, but were connected somehow to navigable waters or "adjacent" to those waters. (The act requires a federal permit to discharge dredged or fill materials into "navigable waters.") Many legal and other observers hoped that the Court's ruling in these cases would bring greater clarity about the scope of federal jurisdiction.

---

[9] Scientists generally agree that the presence of a wetland can be determined by a combination of soils, plants, and hydrology. See the discussion in CRS Report RL33483, *Wetlands: An Overview of Issues*.

[10] 531 U.S. 159 (2001).

The Court's ruling on the two cases was issued in June 2006.[11] In a 5-4 decision, a plurality of the Court, led by Justice Scalia, held that the lower court had applied an incorrect standard to determine whether the wetlands at issue are covered by the CWA. Justice Kennedy joined this plurality to vacate the lower court decisions and remand the cases for further consideration, but he took different positions on most of the substantive issues raised by the cases, as did four other dissenting Justices.[12] Because the several opinions written by the Justices did not draw a clear line regarding which wetlands and other waters are subject to federal jurisdiction, one result has been more case-by-case determinations and continuing litigation. There also has been pressure on the Corps and EPA to clarify the issues through an administrative rulemaking.

The full extent of impacts on the regulatory program resulting from these decisions still remains unclear, in part because of different interpretations of both rulings reflected in subsequent federal court cases. While it continues to be difficult to fully assess how regulatory protection of wetlands has been affected as a result of the decisions and other possible changes, the remaining responsibility to protect affected wetlands falls on states and localities. Environmentalists believe that the Court misinterpreted congressional intent on the matter, while industry and landowner groups welcomed the rulings. Policy implications of how much the decisions restrict federal regulation depend on how broadly or narrowly the opinions are applied. Some federal courts have interpreted *SWANCC* and *Rapanos* narrowly, thus limiting effects on existing permit rules, while a few have read the decisions more broadly, resulting in a more restrictive interpretation of regulatory jurisdiction.

Corps/EPA Guidance

Following both the *SWANCC* and *Rapanos* rulings, EPA and the Corps issued guidance in 2003 and 2008 to enable their field staffs to make CWA jurisdictional determinations in light of the decisions. Environmental

---

[11] Rapanos v. United States, 547 U.S. 715 (2006).

[12] For additional information, see CRS Report RL33263, *The Wetlands Coverage of the Clean Water Act (CWA): Rapanos and Beyond*, by Robert Meltz and Claudia Copeland.

groups criticized the guidance, saying that the agencies are substantially limiting the scope of waters that are protected by the CWA. Industry groups such as developers remain frustrated by what they see as inconsistencies and delays in obtaining needed permits.

The Obama Administration entered this debate in 2011, when EPA and the Corps proposed new guidance, which was intended to clarify regulatory jurisdiction over U.S. waters and wetlands, consistent with the Supreme Court decisions and agency regulations. Like previous guidance documents, the Obama draft examined current regulatory definitions of waters that are subject to CWA jurisdiction, such as interstate waters, and tributaries (at 33 C.F.R. §328.3 and 40 C.F.R. §230.3) in light of the Supreme Court's rulings to determine which waters are clearly subject to the CWA, which waters are not, and which waters require a case-specific analysis in order to determine jurisdiction.

The 2011 proposed guidance quickly generated substantial controversy. Some critics argued that the guidance represented over-reaching by the agencies, beyond authority provided by Congress. Others faulted the continued reliance on federal guidance, which is not binding and lacks the force of law, yet can have significant impact on regulated entities. For various reasons, the 2011 draft guidance was not finalized, and in September 2013, EPA and the Corps announced that the document had been withdrawn from interagency review and also announced that revised regulations to define "waters of the United States" were being developed. On March 25, 2014, the agencies released a proposed rule, and accepted public comment on it until November 14, 2014.[13]

According to EPA and the Corps, the agencies' intent was to clarify CWA jurisdiction, not expand it. Nevertheless, the rule has been extremely controversial, especially with groups representing property owners, land developers, and the agriculture sector, who contend that it represents a massive federal overreach beyond the agencies' statutory authority. Most state and local officials are supportive of clarifying the extent of CWA-

[13] Department of Defense, Department of the Army, Corps of Engineers, and Environmental Protection Agency, "Definition of 'Waters of the United States' Under the Clean Water Act, Proposed Rule," 79 *Federal Register* 22188- 22274, April 21, 2014.

regulated waters, but some are concerned that the rule could impose costs on states and localities as their own actions (e.g., transportation or public infrastructure projects) become subject to new requirements. Most environmental advocacy groups welcomed the intent of the proposal to more clearly define U.S. waters that are subject to CWA protections, but beyond that general support, some favored even a stronger rule.

2015 Revised Rule

On May 27, 2015, EPA and the Corpson Agency (EPA) issued a final rule revising their regulations that define the scope of waters protected under the CWA. The revised rule became effective on August 28, 2015, 60 days after publication in the *Federal Register*.[14]

The revised rule retains much of the structure of the agencies' existing definition of "waters of the United States."[15] It focuses particularly on clarifying the regulatory status of surface waters located in isolated places in a landscape and streams that flow only part of the year, along with nearby wetlands—the types of waters with ambiguous jurisdictional status following the Supreme Court's rulings. Like the 2003 and 2008 guidance documents and the 2014 proposal, it identifies categories of waters that are and are not jurisdictional, as well as categories of waters and wetlands that require a case-specific evaluation.

- Under the final rule, all tributaries to the nation's traditional navigable waters, interstate waters, the territorial seas, or impoundments of these waters would be jurisdictional per se. All of these waters are jurisdictional under existing rules, but the term "tributary" is newly defined in the rule.
- Waters—including wetlands, ponds, lakes, oxbows, and similar waters—that are adjacent to traditional navigable waters, interstate

---

[14] Department of the Army, Corps of Engineers, and Environmental Protection Agency, "Clean Water Rule: Definition of 'Waters of the United States,' Final Rule," 80 *Federal Register* 37054-37127, June 29, 2015.

[15] The definition of "waters of the United States" is found at 33 C.F.R. §328.3 (Corps) and 40 C.F.R. §122.2 (EPA). The term is similarly defined in other EPA regulations, as is the term "navigable waters."

waters, the territorial seas, jurisdictional tributaries, or impoundments of these waters would be jurisdictional by rule. The final rule for the first time puts some boundaries on what is considered "adjacent."

- Some waters—but fewer than under current practice—would remain subject to a case-specific evaluation of whether or not they meet the legal standards for federal jurisdiction established by the Supreme Court. The final rule establishes two defined sets of additional waters that will be a "water of the United States" if they are determined to have a significant nexus to a jurisdictional waters.
- The final rule identifies a number of types of waters to be excluded from CWA jurisdiction. Some are restatements of exclusions under current rules (e.g., prior converted cropland); some have been excluded by practice and would be expressly excluded by rule for the first time (e.g., groundwater, some ditches). Some exclusions were added to the final rule based on public comments (e.g., stormwater management systems and groundwater recharge basins). The rule makes no change and does not affect existing statutory exclusions: permit exemptions for normal farming, ranching, and silviculture practice and for maintenance of drainage ditches (CWA §404(f)(1)), as well as for agricultural stormwater discharges and irrigation return flows (CWA §402(l)).

The agencies' intention was to clarify questions of CWA jurisdiction, in view of the Supreme Court's rulings and consistent with the agencies' scientific and technical expertise. Much of the controversy since the Court's rulings has centered on the many instances that have required applicants for CWA permits to seek a time-consuming case-specific evaluation to determine if CWA jurisdiction applies to their activity, due to uncertainty over the geographic scope of the act. In the rule, the Corps and EPA intended to clarify jurisdictional questions by clearly articulating categories of waters that are and are not protected by the CWA and thus limiting the types of waters that still require case-specific analysis.

However, critical response to the proposal from industry, agriculture, many states, and some local governments was that the rule was vague and ambiguous and could be interpreted to enlarge the regulatory jurisdiction of the CWA beyond what the statute and the courts allow.

Officials of the Corps and EPA vigorously defended the proposed rule. But they acknowledged that it raised questions that required clarification in the final rule. In an April 2015 blog post, the EPA Administrator and the Assistant Secretary for the Army said that the agencies responded to criticisms of the proposal with changes in the final rule, which was then undergoing interagency review. The blog post said that the final rule would make changes such as: defining tributaries more clearly; better defining how protected waters are significant; limiting protection of ditches to those that function like tributaries and can carry pollution downstream; and preserving CWA exclusions and exemptions for agriculture.[16] The final rule announced on May 27 does reflect a number of changes from the proposal, especially to provide more bright line boundaries and simplify definitions that identify waters that are protected under the CWA.[17] The agencies' intention has been to clarify the rules and make jurisdictional determinations more predictable, less ambiguous, and more timely. Based on press reports of stakeholders' reactions to the final rule, it appears that some believe that the agencies largely succeeded in that objective, while others believe that they did not.[18]

Legal challenges to the Clean Water Rule were filed in multiple federal courts soon after it was announced. These lawsuits, filed by industry groups, more than half of the states, and several environmental groups (nearly 90 plaintiffs so far), will test whether the agencies' interpretation of CWA jurisdiction is consistent with the Supreme Court's rulings and whether the rule complies with substantive and procedural requirements of the CWA and other laws.

---

[16] Gina McCarthy and Jo-Ellen Darcy, "Your Input Is Shaping the Clean Water Rule," *EPA Connect, The Official Blog of EPA's Leadership*, April 6, 2015, http://blog.epa.gov/epaconnect/2015/04/your-input-is-shaping-the-clean-waterrule/#more-3470.

[17] See CRS Report R43455, EPA and the Army Corps' Rule to Define "Waters of the United States" for discussion.

[18] See, for example, Amena H. Saiyid, "Obama Says Water Jurisdiction Rule Provides Clarity, Certainty; Critics Claim Overreach," *Daily Environment Report*, May 28, 2015, pp. A-1.

Because of uncertainty about the correct judicial venue for challenging the rule,[19] petitions for review have been filed both in federal district courts and courts of appeal. As of December 1, 2015, petitions for review of the rule have been filed in eight appellate courts; most have been consolidated in the Sixth Circuit. Sixteen separate challenges also were filed in 12 federal district courts. On October 9, a three-judge panel of the U.S. Court of Appeals for the Sixth Circuit placed a nationwide stay on the 2015 rule, pending further developments, including the need to determine the court's own jurisdictional authority.[20] On the substance of the complaints, the court said there was a good chance that the plaintiffs would prevail on the merits. A two-judge majority said that the significance of the new rule warranted leaving the prior regulatory regime in place, while the third judge said that until the question of subject-matter jurisdiction is answered, the new rule should not be stayed.[21] The Sixth Circuit court heard arguments on December 8, 2015, on whether it has exclusive jurisdiction to review the rule, but the court has not yet issued its ruling. As a result of the court's order, the Corps and EPA will continue to make CWA jurisdictional determinations based on the 2008 guidance, as they did before promulgation of the 2015 rule.

Legislative Responses

Congressional committees have held oversight hearings on both the *SWANCC* and *Rapanos* decisions, seeking clarification of interpretations

---

[19] The judicial review section of the CWA, Section 509, vests exclusive, original review jurisdiction over enumerated EPA actions under the act in the federal courts of appeals. The initial issue with Section 509 is that none of the listed EPA actions clearly cover the Clean Water Rule. Indeed, in the preamble to the final rule, EPA and the Corps acknowledge that "[t]he Supreme Court and lower courts have reached different conclusions on the types of actions that fall within section 509," and offers no opinion of its own as to review of the Clean Water Rule. If a court finds that the rule is not covered by Section 509, review jurisdiction presumably will lie in the district courts pursuant to the federal question statute. That statute, applicable where no more specific statute provides otherwise, gives the district courts original jurisdiction over "all civil actions arising under the ... laws ... of the United States." (28 U.S.C. §1331) See CRS Legal Sidebar WSLG1369, *The EPA/Corps Clean Water Rule: What Court or Courts Get to Rule on the Legal Challenges?*

[20] On August 27, a district court in North Dakota issued a preliminary injunction that blocked implementation of the rule in 13 states, but not in the remaining 37 states.

[21] In re Environmental Protection Agency and Department of Defense, Nos. 15-3799 et al. (6th Cir., Oct. 9, 2015), http://www.ca6.uscourts.gov/opions.pdf/15a0246p-06.pdf.

and impacts of the rulings. But the uncertainties about federal jurisdiction over wetlands and other waters raised by the rulings remain highly controversial. In response, legislation to overturn the decisions by providing a broad definition of "waters of the United States" has been introduced regularly since the 107th Congress, and one such bill was reported by a Senate committee in the 111th Congress.[22] Legislation that instead would narrow the definition of "waters of the United States" also has been introduced.

Environmental advocates and others contend that Congress must clarify the important issues left unsettled by the Supreme Court's 2001 and 2006 rulings and by the Corps/EPA guidance. They also argue that legislation is needed to "reaffirm" what Congress intended when the CWA was enacted in 1972 and what EPA and the Corps have subsequently been practicing until the two Supreme Court rulings, in terms of CWA jurisdiction. But critics questioned the constitutionality of legislation that was proposed and asserted that it would expand federal authority, thus likely increasing confusion, rather than settling it.

EPA's and the Corps' efforts to develop revised *Rapanos* guidance and revised regulations have been controversial and received congressional attention. Legislative provisions to prohibit the agencies from funding activities related to the guidance and the proposed "waters" rule were included in appropriations bills since the 112th Congress. One such provision was enacted as part of the Consolidated and Further Continuing Appropriations Act, 2015 (P.L. 113-235) in December 2014. It required EPA and the Corps to withdraw an interpretive rule related to the proposed "waters" rule, which addressed permit exemptions for agricultural activities but created controversy and confusion.[23]

Congressional interest has continued to be strong in the 114th Congress. On February 4, 2015, the Senate Environment and Public Works Committee and the House Transportation and Infrastructure Committee held a joint hearing on impacts of the 2014 proposed rule on state and local

---

[22] For information, see CRS Report RL33263, The Wetlands Coverage of the Clean Water Act (CWA): Rapanos and Beyond.

[23] See CRS Insight IN10212, Withdrawal of the EPA-Army Corps Interpretive Rule for Agriculture, by Claudia Copeland.

governments, hearing from public and EPA and Corps witnesses. Hearings also have been held by other congressional committees. As well, a number of bills have been introduced, most of them intended either to prohibit the agencies from finalizing the 2014 proposed rule or to detail procedures for a new rulemaking. Bills have been of several types.[24]

- Joint resolutions of disapproval under the Congressional Review Act. The Senate has passed such a resolution (S.J.Res. 22), and a similar resolution has been introduced in the House (H.J.Res 59).
- Appropriations bill limitations. Bills with limitations were reported in the Senate and House in 2015, but the FY2016 Consolidated Appropriations Act (P.L. 114- 113) contained no such provisions.
- Standalone targeted legislation. The House has passed one such bill (H.R. 1732). Similar legislation was reported in the Senate, but failed to advance (S. 1140).
- Broad amendments to the CWA to affirm or clarify Congress's intention regarding CWA jurisdiction, including such bills as S. 980 and S. 2705.

None of these bills has been enacted, but congressional interest in halting or modifying the "waters of the United States" rule is likely to remain high in 2016.

Authorization of Clean Water Infrastructure Funding

Meeting the nation's needs to build, upgrade, rebuild, and repair wastewater infrastructure is a significant element in achieving the CWA's water quality objectives and an issue of continuing interest to policymakers. The act's program of financial aid for municipal wastewater treatment plant construction is a key contributor to that effort. Since 1972, Congress has provided nearly $94 billion to assist local governments in

[24] For additional discussion, see CRS Report R43943, EPA and the Army Corps' "Waters of the United States" Rule: Congressional Response and Options, by Claudia Copeland.

constructing projects to achieve the act's requirements for secondary treatment of municipal sewage (equivalent to 85% reduction of wastes), or more stringent treatment where required by local water quality conditions. State and local governments have spent more than $25 billion of their own funds for construction, as well. Federal funds can only be used for construction purposes (i.e., new plants or upgrades), but not for operation and maintenance of facilities, which are funded from local sources.

Still, funding needs remain very high: an additional $298 billion, according to the most recent Needs Survey estimate by EPA and the states, a 17% increase above the estimate reported four years earlier.[25] This estimate includes $187.9 billion for wastewater treatment and collection systems ($26.7 billion more than the previous report), which represent more than 60% of all needs; $63.6 billion for combined sewer overflow corrections ($1.4 billion less than the previous estimate); $42.3 billion for stormwater management ($17 billion more than the previous estimate); and $4.4 billion to build systems to distribute recycled water ($700 million less than the previous estimate).

While water infrastructure investments are made and projects are built, new funding needs also are identified. Rather than decreasing over time, estimates of funding needs continue to grow. EPA reported several reasons for increased funding needs, which were $23 billion higher than in the previous report. Cited reasons include improvements needed to meet more protective water quality standards, rehabilitation of aging infrastructure, and expanding capacity to meet population growth. Needs for stormwater management increased by $17 billion and were mostly due to emerging needs to provide "green" infrastructure (e.g., use of wetland and other natural systems to capture stormwater) as a supplement to traditional stormwater treatment structures. The estimates do not explicitly include funding needed to address security issues, or funding possibly needed for treatment works to adapt to climate change impacts.

[25] U.S. Environmental Protection Agency, *Clean Watersheds Needs Survey 2008, Report to Congress*, Washington, June 2010, http://water.epa.gov/scitech/datait/databases/cwns/upload/cwns2008rtc.pdf.

Debate over the nation's efforts regarding wastewater infrastructure was a central part of the 1987 CWA amendments. The amendments extended through FY1990 the traditional Title II program of grants for sewage treatment project construction, under which the federal share was 55% of project costs. The 1987 law initiated a program of grants to capitalize State Water Pollution Control Revolving Funds (SRFs), which are loan programs, in a new Title VI. Under the revolving fund concept, monies used for wastewater treatment construction are repaid by loan recipients to the states (repayment was not required for grants under the Title II program), to be used for future construction in other communities, thus providing an ongoing source of financing. The expectation in 1987 was that the federal contributions to SRFs would assist in making a transition to full state and local financing by FY1995. Although most states believe that the SRF is working well, continuing large funding needs have delayed the anticipated shift to full state responsibility. Thus, SRF issues have been prominent on the Clean Water Act reauthorization agenda in recent Congresses.[26]

SRF monies may be used for specified activities, including making loans for as much as 100% of project costs (at or below market interest rates, including interest-free loans), to buy or refinance cities' debt obligation, or as a source of revenue or security for payment of principal and interest on a state-issued bond. SRF monies also may be used to provide loan guarantees or credit enhancement for localities. Loans made by a state from its SRF are to be used first to assure progress toward the goals of the act and, in particular, on projects to meet the standards and enforceable requirements of the act. After states achieve those requirements, SRF monies also may be used to implement national estuary programs and nonpoint pollution management. Since the SRF program began, states have used about 4% of clean water SRF funds to assist nonpoint management projects and estuary projects.

All states have established the mechanisms to administer the loan program and have been receiving SRF capitalization funds under Title VI.

---

[26] For further information on the clean water SRF program, see CRS Report 98-323, *Wastewater Treatment: Overview and Background*, by Claudia Copeland.

Congressional oversight has examined the progress toward reducing the backlog of wastewater treatment facilities needed to achieve the act's water quality objectives, while estimates of future funding needs have drawn increased attention to the role of the SRF program in meeting such needs. Although there has been some criticism of the SRF program, and debate continues over specific concerns, the basic approach is well supported. Congress used the clean water SRF as the model when it established a drinking water SRF in 1996 (P.L. 104-182).[27]

The initial intent was to phase out federal support for this program, but Congress has continued to appropriate SRF capitalization grants to the states—a total of $41 billion since the 1987 amendments, providing an average of $1.45 billion annually in recent years. *Table 1* summarizes recent Administration budget requests and enacted appropriations for SRF capitalization grants. This table does not include appropriations for congressionally directed special project grants in individual cities (that is, congressional earmarks), which for several years represented about 15% of water infrastructure funds.[28]

One issue of continuing interest is impacts of paying for water infrastructure projects on small communities, many of which have found it difficult to participate in the SRF loan program. This is due to a number of factors. Many are characterized by narrow or weak tax bases, limited or no access to capital markets, lower relative household incomes, higher per capita needs, and limited ability to demonstrate economies of scale. They often find it harder to borrow to meet their capital needs and pay relatively high premiums to do so. Meeting the special needs of small towns, through a reestablished grant program, other funding source, or loan program with special rules, has been an issue of interest to Congress.

---

[27] For additional information, see CRS Report RS22037, *Drinking Water State Revolving Fund (DWSRF): Program Overview and Issues*, by Mary Tiemann.

[28] Issues associated with special project grants are discussed in CRS Report RL32201, *Water Infrastructure Projects Designated in EPA Appropriations: Trends and Policy Implications*, by Claudia Copeland. Since FY2011, Congress has placed a moratorium on earmarks, but some policymakers favor restoring the practice.

Table 1. Clean Water SRF Capitalization Grants, FY2008-FY2016

(millions of dollars)

Fiscal Year	President's Request	Appropriations
2008	687.6	1,083.8
2009	555.0	4,689.1[a]
2010	2,400.0	2,100.0
2011	2,000.0	1,522.0
2012	1,550.0	1,466.5
2013	1,175.0	1,376.1[b]
2014	1,095.0	1,448.9
2015	1,018.0	1,448.9
2016	1,116.0	1,393.9
TOTAL	*11,596.6*	*16,529.2*

Source: Compiled by CRS.

[a] FY2009 appropriations include $4.0 billion in supplemental appropriations as part of the American Recovery and Reinvestment Act of 2009 (P.L. 111-5).

[b] FY2013 appropriations reflect post-sequester/post-rescission amount.

Because remaining clean water funding needs are still so large nationally, at issue is whether and how to extend SRF assistance to address those needs, how to allocate SRF funds among the states, and how to prioritize projects and funding. Additionally, there is concern about the adequacy of SRF or other funding specifically for high-cost projects dealing with problems of overflows from municipal combined and separate sewers which can release partially treated or untreated wastewaters that harm public health and the environment. EPA estimates that the cost of projects to control sewer overflows and manage stormwater runoff is nearly $64 billion nationwide—more than one-fifth of all needs estimated in the most recent Needs Survey. Wastewater utilities also have sought assistance to assess operational vulnerabilities and upgrade physical protection of their facilities against possible terrorist attacks that could threaten the water infrastructure system.[29]

[29] For additional information on many of these topics, see CRS Report RL31116, *Water Infrastructure Needs and Investment: Review and Analysis of Key Issues*, by Claudia Copeland and Mary Tiemann.

In 2010 EPA issued a "Clean Water and Drinking Water Infrastructure Sustainability Policy" addressing management and pricing of infrastructure funded through SRFs to encourage conservation and provide adequate long-term funding for future capital needs. EPA is working with water utilities to promote planning processes that reflect not only public health and water quality, but also conservation of natural resources and innovative treatment. Further, EPA is working with states to target SRF assistance to projects that focus on system upgrade and replacement in existing communities, reflect full life cycle costs of infrastructure assets, and conserve natural resources or use alternative approaches.

Issues Affecting Legislative Efforts

Congress had considered water infrastructure funding issues several times since the 107th Congress, but no legislation other than appropriations was enacted until P.L. 113-121, discussed next. Despite specific issues that have stalled legislation, the act's water infrastructure program is widely supported both inside and outside Congress. However, because the House and Senate have focused extensively on reducing federal spending and deficit reduction recently, proposals concerning new or expanded federal spending for water infrastructure investments have not advanced. Throughout this period, several factors contributed to difficulties in moving bills through the legislative process. They included Bush Administration opposition to higher authorization levels, controversies over application of prevailing wage requirements of the Davis-Bacon Act to water infrastructure projects, and disputes over the formula for allocating clean water SRF grants among the states.

The issue of the applicability of the Davis-Bacon Act to SRF-funded projects has been especially controversial, because that act has both strong supporters and critics in Congress and elsewhere. It requires, among other things, that not less than the locally prevailing wage be paid to workers employed, under contract, on federal construction work "to which the United States or the District of Columbia is a party." Critics of Davis-Bacon say that it unnecessarily increases public construction costs and hampers competition, while supporters say that it helps stabilize the local construction industry by preventing competition that would undercut local wages and working conditions. Under the original SRF program authorization enacted in 1987, the Davis-Bacon Act applied to so-called "first use" monies provided by a state from its SRF (that is, loans made from initial federal capitalization grants, but not to subsequent monies provided from repayments to the SRF). When that authorization expired at the end of FY1994, Davis-Bacon requirements also expired. Thus, the recent issue has been whether to restore the applicability of those requirements.[30] While authorizing committees have debated this issue for some time, Davis-Bacon requirements have been attached to use of SRF funds through appropriations acts since 2009.

A second issue that has complicated enactment of legislation is the method of allocating SRF capitalization grants among the states. CWA Section 205(c)(3) contains a table that identifies each state's percentage share of appropriated funds. Changing the formulation of how funds are distributed matters to every state, because inevitably it results in "winners" and "losers." But because the existing statutory allotment has not been revised since 1987, while needs have changed considerably, the issue is important to considering clean water infrastructure legislation.[31]

WIFIA Pilot Program and SRF Amendments in P.L. 113-121

Most policymakers acknowledge that communities face formidable challenges in providing and paying for adequate and reliable water infrastructure services for their citizens, and Congress has long considered ways to help meet those challenges. Several policy options have been discussed, including some that exist and are well established—such as the SRF program—while some are newer—such as creating a national infrastructure bank. Some are intended to provide long-term revenue to support infrastructure financing programs, and some are intended to

---

[30] For information, see CRS Report R41469, Davis-Bacon Prevailing Wages and State Revolving Loan Programs Under the Clean Water Act and the Safe Drinking Water Act, by Gerald Mayer and Jon O. Shimabukuro.

[31] For additional information on the current statutory formula, see CRS Report RL31073, Allocation of Wastewater Treatment Assistance: Formula and Other Changes, by Claudia Copeland.

encourage private participation in providing wastewater services. At this point, there is no consensus favoring a single policy, and many advocate a combination of options to expand the financing "toolbox."

One particular option that has been debated is a "Water Infrastructure Finance and Innovation Act," or WIFIA, program, and legislation to create a WIFIA pilot program was enacted in the 113th Congress (H.R. 3080/P.L. 113-121). The legislation, the Water Resources Reform and Development Act (WRRDA) Title V, Subtitle C, authorizes a five-year WIFIA pilot program. Under the bill, EPA is authorized to provide credit assistance (secured loans or loan guarantees) for drinking water and wastewater projects, and the U.S. Army Corps of Engineers is authorized to provide similar assistance for water resource projects, such as flood control or hurricane and storm damage reduction. Proponents argue that WIFIA offers a number of financing advantages, such as providing credit assistance at low U.S. Treasury rates to projects that otherwise have difficulty obtaining financing, thus lowering the cost of capital to borrowers. Under the legislation, EPA and the Corps each are authorized a total of $175 million over five years (beginning with $20 million in FY2015 and increasing to $50 million in FY2019 for each agency) to provide assistance. Projects must be $20 million or larger in costs to be eligible, except that projects in rural areas (population 25,000 or less) must have eligible projects costs of $5 million or more.[32]

Since passage of the legislation, EPA has held a series of meetings around the country to discuss implementation of the new program, as the agency seeks stakeholder views on issues such as how to define eligible projects, develop a project ranking system and evaluation criteria, and determine credit worthiness. However, the program will not proceed until Congress provides appropriations for it. In the FY2015 omnibus appropriations act, enacted in December 2014 (P.L. 113-235), Congress provided EPA with $2.2 million for hiring and staffing to implement the new program, but it did not appropriate funds to actually finance projects. The Consolidated Appropriations Act, 2016, enacted in December 2015

---

[32] For additional information, see CRS Report R43315, Water Infrastructure Financing: The Water Infrastructure Finance and Innovation Act (WIFIA) Program, by Claudia Copeland.

(P.L. 114-113) provided EPA with $5.0 million to continue preparation for the program. Neither bill appropriated funds to actually finance projects.

In addition to the WIFIA provisions, P.L. 113-121 includes a number of provisions amending certain water infrastructure provisions of the CWA, especially the Title VI SRF program. Some of the provisions in P.L. 113-121 were included in other legislation and proposals in recent Congresses that have not advanced (such as extending SRF loan repayment from 20 to 30 years, allowing states to make subsidized loans under certain circumstances, adding land acquisition to the definition of "treatment works" in order to be eligible for SRF assistance, and explicitly allowing SRF monies to be used for security projects at wastewater treatment plants). Several of the provisions have been included in recent appropriations bills and are now codified in the CWA by P.L. 113-121 (such as expanding the list of SRF-eligible projects to include energy- and water-efficiency and others; increasing SRF assistance to Indian Tribes; and imposing "Buy American" requirements on SRF assistance). The CWA provisions included in P.L. 113-121 are the first amendments to Title VI since 1987. However, the amendments do not address other long-standing Title VI issues: authorization of appropriations for capitalization grants (i.e., the amendments do not reauthorize clean water SRF grants), state-by-state allocation of capitalization grants (i.e., the allocation formula that has been in effect since 1987), or applicability of prevailing wage requirements under the Davis-Bacon Act (locally prevailing wages are to be paid to workers on projects that receive SRF assistance).[33]

Other Clean Water Act Issues

A number of other issues affecting efforts to achieve the goals and objectives of the CWA have drawn interest recently and been the subject of congressional oversight and legislation. Some legislators have been highly

[33] For additional information on P.L. 113-121, see CRS Report R43298, *Water Resources Reform and Development Act of 2014: Comparison of Select Provisions*, by Nicole T. Carter et al.

critical of recent regulatory initiatives, while others have been more supportive of EPA's implementation efforts.

Since 2009, EPA has proposed and promulgated numerous regulations implementing the CWA and other pollution control statutes that it administers. Critics of the Administration, both within Congress and outside of it, have accused the agency of reaching beyond the authority given it by Congress and ignoring or underestimating the costs and economic impacts of these rules. Majority party leaders in the House conducted vigorous oversight of the agency in the 112th and 113th Congresses. Bills seeking to overturn specific regulations or to limit the agency's authority also were introduced, along with proposals to bar EPA funding for specific activities. Environmental groups disagree that the agency has overreached, and EPA itself contends that critics' focus on the cost of controls obscures the benefits of new regulations. The agency estimates that benefits far exceed the costs, and that investing in pollution control is an important source of economic activity, exports, and American jobs. Although particular attention is being paid to the Clean Air Act, a number of EPA's initiatives concerning the CWA also have received legislators' scrutiny.[34] In several cases, policymakers have sought to curtail water quality protection initiatives under the CWA following court rulings that expanded the regulatory scope of the law. In the 114th Congress, scrutiny of EPA initiatives has continued to be intense, including those involving water quality.

Mountaintop Mining in Appalachia

One water quality issue that has received considerable attention is mountaintop coal mining. Mountaintop removal coal mining involves removing the top of a mountain in order to recover the coal seams contained there. This practice occurs in six Appalachian states (Kentucky, West Virginia, Virginia, Tennessee, Pennsylvania, and Ohio). It creates an immense quantity of excess spoil, which is typically placed in nearby valleys, burying streams that flow through the valleys. Critics say that, as a

---

[34] For information, see CRS Report R41561, *EPA Regulations: Too Much, Too Little, or On Track?*, by James E. McCarthy and Claudia Copeland.

result of valley fills, stream water quality and the aquatic and wildlife habitat that streams support are destroyed. The mining industry argues that mountaintop mining is essential to conducting surface coal mining in the Appalachian region and that surface coal mining would not be economically feasible there if producers were restricted from using valleys for the disposal of mining overburden.[35]

Mountaintop mining is regulated under several laws, including the CWA Section 404 permit program (discussed above) and the Surface Mining Control and Reclamation Act. In June 2009, officials of EPA, the Corps of Engineers, and the Department of the Interior's Office of Surface Mining and Reclamation (OSM) signed a memorandum of understanding outlining a series of administrative actions under these laws to reduce the harmful environmental impacts of mountaintop mining and surface coal mining in Appalachia. The plan includes a series of near-term and longer-term actions that emphasize specific steps, improved coordination, and greater transparency of decisions. The actions are being implemented through regulatory proposals, guidance documents, and review of pending applications for permits to authorize mountaintop mining-valley fill operations. In 2009, the Army Corps suspended the use of a particular CWA general permit (nationwide permit 21) for surface coal mining activities in Appalachia and in 2012 it finalized rules to apply more stringent use of CWA general permits by these coal mining operations.[36]

Also in 2009 EPA and the Corps began conducting detailed evaluations of 79 pending CWA permit applications for surface mining activities in order to limit environmental impacts of the proposed activities under a process called Enhanced Coordination Procedures (ECP). Coal industry groups and coal state officials contended that the ECP process resulted in costly delay in issuance of permits. They challenged the process in federal court, and in October 2011, the court struck down the ECP as an

---

[35] For additional information, see CRS Report RS21421, *Mountaintop Mining: Background on Current Controversies*, by Claudia Copeland.

[36] For information, see CRS Report 97-223, *The Army Corps of Engineers' Nationwide Permits Program: Issues and Regulatory Developments*, by Claudia Copeland.

unlawful transfer of legal authority from the Corps to EPA.[37] Thereafter, the agencies continued to review permit applications for surface coal mining projects in Appalachia under existing rules, but not the vacated ECP.

In 2011 EPA issued guidance on review of CWA Section 402 and 404 permit requests for surface coal mining in Appalachia. The guidance tightened oversight of permit reviews in several ways, most notably by establishing two benchmarks for stream conductivity, which is a measure of the level of salinity in water and is a proxy for dissolved solids in stream waters associated with mining activity that may contribute to toxicity. The guidance has been very controversial with industry. The House Transportation Subcommittee on Water Resources and Environment held hearings on these issues in May 2011. A hearing also was held by the House Government Reform and Oversight Committee in July 2011. In July 2012, the same federal court that struck down the ECP also invalidated the 2011 guidance document intended to help assess a mine's water quality impacts, ruling that EPA had overstepped its statutory authority. The government appealed both of these rulings, which were overturned by a federal appeals court in July 2014.[38]

Legislation to restrict the practice of mountaintop mining has been introduced. A bill in the 114th Congress, H.R. 912, would place a moratorium on permitting for mountain removal coal mining until certain health studies are conducted. A bill in the 113th Congress (H.R. 1837, the Clean Water Protection Act) would have narrowed the CWA definition of "fill material," and thus narrowed the types of materials that can be discharged into U.S. waters under a Section 404 permit. The significance of the bill is that discharges of materials that are not eligible for a Section 404 permit are regulated under CWA Section 402. Because Section 402 discharge requirements are more restrictive than those for Section 404, some discharges that could be permitted under Section 404 cannot be authorized under Section 402. Supporters favored making it more difficult

---

[37] At the time of the court's ruling, 8 of the 79 projects under ECP review had received permits; 50 permit applications had been withdrawn by the applicants; 3 project reviews were underway or nearly complete; and 18 reviews had not yet begun.

[38] National Mining Association v. McCarthy, D.C. Cir., No. 12-5310, July 11, 2014.

to use Section 404 to authorize activities that they consider to be environmentally harmful. On the other hand, critics of the legislation say that, as a practical matter, economically important activities such as coal mining could not meet the more stringent limitations of a Section 402 permit and, thus, would be infeasible.[39]

Another aspect of the mountaintop mining issue that has drawn attention is EPA's 2011 veto of a CWA Section 404 permit for a surface coal mining operation in West Virginia, the Spruce No. 1 mine. EPA's action has been controversial, particularly because the veto occurred after the permit had been issued by the Army Corps. EPA's veto of the permit was challenged, and in 2012, a federal district court overturned the veto, ruling that EPA had exceeded its statutory authority in the Spruce No. 1 action. However, in 2013, a federal appeals court reversed the lower court's decision and upheld EPA's authority to retroactively veto permits.[40] The appeals court ruling was applauded by environmental groups and criticized by the mining industry. In response, bills have been introduced to limit or prohibit EPA's ability to exercise this veto authority, contained in CWA Section 404(c). Several proposals in the 114th Congress (S. 55/S. 234, H.R. 896, and H.R. 1203) would bar EPA from vetoing a 404 permit retroactively.

CONTINUING ISSUE: APPROPRIATIONS

Although few CWA amendments have been enacted recently, clean water policy and program issues have been addressed regularly by Congress in the context of annual appropriations acts.

EPA's appropriations are included in the Interior, Environment, and Related Agencies appropriations acts.[41]

---

[39] For additional information, see CRS Report RL31411, *Controversies over Redefining "Fill Material" Under the Clean Water Act*, by Claudia Copeland.

[40] Mingo Logan Coal Company v. U.S. Environmental Protection Agency, 714 F.3d 608 (DC Cir. 2013); *cert. denied*, 134 S. Ct. 1540 (2014). For background on the veto, see CRS Report RS21421, *Mountaintop Mining: Background on Current Controversies*, by Claudia Copeland.

[41] For additional information, see CRS Report 96-647, *Water Infrastructure Financing: History of EPA Appropriations*, by Claudia Copeland.

FY2016 Appropriations

The Administration's FY2016 budget was presented on February 2, 2015. Overall, the budget sought $8.6 billion for EPA, or $452 million more than the FY2015 enacted level of funding.[42] The EPA request included $1.116 billion for clean water SRF capitalization grants ($333 million less than the FY2015 enacted level; see *Table 1*) and $15 million for Alaska Native Village and U.S.-Mexico Border water infrastructure projects. The Administration sought $238.8 million to protect surface waters (19.5% higher than the FY2015 enacted level), and $250 million for the Great Lakes Restoration Initiative (17% below the FY2015 enacted level). The budget sought increases for several water quality grant programs (Section 106 grants, Section 319 grants, and wetlands program development grants.

Although the House and Senate Appropriations Committees reported bills to provide FY2016 appropriations for EPA (H.R. 2822 and S. 1645), final appropriations action for EPA and other agencies occurred as part of the Consolidated Appropriations Act, 2016, signed by the President December 18, 2015 (P.L. 114-113). The bill provides $1.393.9 billion for clean water SRF capitalization grants ($55 million less than FY2015, but $278 million above the President's request) and $30 million for Alaska Native Village and U.S.-Mexico Border water infrastructure projects. It also provides $300 million for the Great Lakes Restoration Initiative. The final bill includes the requested increase in funds for Section 319 nonpoint pollution management grants, but not for other water quality grant programs (i.e., Section 106 grants and wetlands program development grants).

---

[42] See CRS Report R44208, *Environmental Protection Agency (EPA): FY2016 Appropriations*, by Robert Esworthy and David M. Bearden.

In: Water Quality
Editors: Edward Saunders et al.
ISBN: 978-1-53614-809-1

Chapter 2

CLEAN WATER ACT SECTION 401: BACKGROUND AND ISSUES*

Claudia Copeland

SUMMARY

Section 401 of the Clean Water Act (CWA) requires that an applicant for a federal license or permit provide a certification that any discharges from the facility will comply with the act, including state-established water quality standard requirements. Disputes have arisen over the states' exercise of this authority in protecting water quality. For the most part, the debate over the Section 401 certification issue has been between states and hydropower interests. A 1994 Supreme Court decision, which upheld the states' authority in this area, dismayed development and hydropower interest groups. The Court revisited these issues in a 2006 ruling that unanimously upheld the authority of states to condition hydropower licenses by exercising Section 401. The dispute between states and industry groups about Section 401 authority has been a legislative issue on several occasions, but Congress has not modified the provision's scope. In the 114th Congress, Senate and House committees

* This is an edited, reformatted and augmented version of Congressional Research Service, Publication No. 97-488, dated July 2, 2015.

are considering a number of energy policy proposals, including legislation that some states believe could impinge on states' authority under CWA Section 401 concerning hydropower projects.

In addition, there has been interest in clarifying whether Section 401 certification applies to nonpoint source discharges, such as rainfall runoff, as well as point source discharges from pipes or ditches. This question was raised in lawsuits in Oregon, where a federal court ruled in 1998 and again in 2008 that Section 401 does not apply to nonpoint source discharges. Still, some interests continue to favor a broad reading of 401 that would apply to both nonpoint and point sources of pollutant discharges.

Background

Under the Clean Water Act (CWA), an applicant for a federal license or permit to conduct any activity that may result in a discharge to waters of the United States must provide the federal agency with a Section 401 certification. The certification, made by the state in which the discharge originates, declares that the discharge will comply with applicable provisions of the act, including water quality standards. A state's water quality standards specify the designated use of a stream or lake (e.g., for water supply or recreation), pollutant limits necessary to protect the designated use (in the form of numeric or narrative criteria), and policies to ensure that existing water uses will not be degraded by pollutant discharges.

Section 401 provides states with two distinct powers: one, the power indirectly to deny federal permits or licenses by withholding certification; and two, the power to impose conditions upon federal permits by placing limitations on certification. Generally, Section 401 certification has been applied to hydropower projects seeking a license from the Federal Energy Regulatory Commission (FERC) and to dredge-and-fill activities in wetlands and other waters that require permits from the U.S. Army Corps of Engineers (CWA Section 404 and Sections 9 and 10 of the Rivers and

Harbors Act). It also is applied to permit requirements for industrial and municipal point source dischargers (CWA Section 402). In addition, it has the potential to be applied to a range of other activities that could affect water quality, a point that has increasingly become an issue.

Because participation by states in Section 401 certification is optional (they may waive the authority if they choose to do so), state implementation has varied. In recent years, however, many states have come to view Section 401 as an important tool in their overall programs to protect the physical and biological, in addition to the chemical, integrity of their waters. Some have begun using Section 401 to address a wide range of impacts to the quality of their waters, including impacts to aquatic habitat such as wetlands where issues of non-chemical impacts arise. Through Section 401, some states have addressed impacts of a project such as inadequate river flow, inundation of habitat, dissolved oxygen levels, and impacts on fish and other wildlife.

This expanded use of Section 401 has, in turn, led to tensions between state and federal agencies (especially FERC) and regulated entities over the scope of the states' Section 401 authority, particularly the extent to which states can legally address water flow requirements in water quality standards. Some state courts have placed limitations on the use of Section 401 (at least for hydropower projects) to address only chemical impacts of projects (such as dissolved oxygen or numeric chemical criteria) and not physical impacts (filling of aquatic habitat in a streambed as a result of the project) or biological impacts (effects on fish migration, for example). Other courts have adopted a broader view and allowed states to condition certification on compliance with all applicable water quality-related laws. A 1990 Supreme Court case[1] addressed the issue of whether hydropower projects must comply with any aspect of state water use law. The Court held that, with regard to federally licensed hydropower facilities, the Federal Power Act preempts state water use law, including states' comprehensive arrangements for allocating water among competing uses.

---

[1] California v. FERC, 495 U.S. 490 (1990), known as the Rock Creek Case.

STAKEHOLDER CONCERNS AND OTHER COURT RULINGS

For the most part, the debate over the Section 401 certification issue has centered on states and hydropower interests. Many states have long favored clarifying the CWA to confirm their broad authority to impose conditions on federally permitted activities. This position was described in testimony at a 1991 Senate subcommittee hearing.[2]

> [A]n overly narrow reading of section 401 would deprive the States of the ability to maintain the very beneficial uses that the Clean Water Act was designed to protect. Federal agencies could permit activities that would undermine a State's investment in pollution control efforts and impose a double standard for different activities affecting the same in-stream values. It makes no sense to authorize States to implement Clean Water Act programs designed to protect beneficial uses and yet leave them powerless to prevent a federally permitted activity from impairing those values.
>
> The comprehensive nature of State management of water quality and water quantity means that the States are best situated to determine whether a federally permitted activity will fully protect beneficial uses. The States have lead responsibility for protecting water quality under the Clean Water Act and for administering laws governing allocation of water quantity. Water quality and quantity are inextricably linked; both are essential to maintaining the integrity of the nation's waters.

Hydropower interests favor allowing federal agencies such as FERC to determine what conditions on a project are necessary for protection of water quality or to satisfy other criteria, in light of the important purposes directed by Congress in other laws, specifically the Federal Power Act.[3]

---

[2] Clive J. Strong, Statement on behalf of the National Association of Attorneys General, in, U.S. Congress, Senate, Committee on Environment and Public Works, Subcommittee on Environmental Protection. Water Pollution Prevention and Control Act of 1991, hearings on S. 1081, 102d Congress, 1st session, Washington: GPO, 1991 (S.Hrg. 102-335), p. 805. (Hereinafter, 1991 Senate Hearing)

[3] Gail Ann Greely, statement on behalf of the National Hydropower Association, in 1991 Senate Hearing, p. 810.

> The current limitation on the role of the States in the [federal hydropower] licensing process is that ultimately the FERC must make the decision balancing the multitude of resource interests affected by the project. The expansive reading of Section 401 water quality certification being used in some States crosses this barrier, using this mandatory water quality review to effectively take control of all aspects of the project.
>
> ... Expansion of 401 certification places authority for an energy resource in the effective control of a State water quality agency, that is not responsible for utility rate stabilization, assuring adequate water supplies, promoting clean air technology, or controlling floods.

The Supreme Court again considered the Section 401 issue in 1994. In *Public Utility District (PUD) No. 1 of Jefferson County and City of Tacoma v. Washington Department of Ecology*,[4] the Court held that a state may impose minimum stream flow requirements as a condition in a Section 401 certification for a proposed hydropower facility because the CWA allows states to condition certification upon any effluent limitation or other appropriate state law requirement, to ensure that the facility will not violate state water quality standards. Imposition of the condition as part of the Section 401 certification does not conflict with FERC's authority to issue a license under the Federal Power Act, the Court said. The ruling said that states may regulate the impacts of a project as a whole, so long as there is a discharge involved. Thus, the conditions a state may require are not confined to the discharge itself but can address a range of conditions as part of their certifications, such as the impacts of a dam's removal of water from a river. Further, federal agencies must include state-imposed conditions in the license or permit.

This decision pleased states, which had sought confirmation of their power to impose minimum stream flow and other requirements of state water quality standards. Environmentalists, who have supported states' use of Section 401 to address aquatic habitat alteration and biological diversity of the nation's waters, were similarly pleased. Development and hydropower interests, on the other hand, said that it would make licensing

---

[4] 511 U.S. 700 (1994).

of hydropower facilities more difficult and costly. The utility industry was concerned that state water quality agencies reflect a narrow viewpoint under their mandates and could bias licensing policies by not adequately addressing power needs.

The Supreme Court revisited these issues in 2006 in a case brought by the owner of several hydropower dams in Maine who had challenged the state's 401 certification for renewal of its FERC licenses for the dams, arguing that the dams did not produce the requisite "discharge" under the CWA. The Court unanimously held that states, through Section 401 certification, can impose conditions on FERC licensing or relicensing of hydropower facilities that states find necessary to prevent adverse alteration of water quality.[5] States and environmental groups applauded the ruling, as many had feared that an adverse decision would hinder the ability of states to require measures to ameliorate the effects of hydropower dams on water quality and aquatic life.

SECTION 401 AND LAND RUNOFF

In 1996, a federal district court in Oregon ruled that Section 401 "applies to all federally permitted activities that may result in a discharge, including discharges from nonpoint sources."[6] The case sought to have the U.S. Forest Service obtain state Section 401 certification that cattle grazing under a Forest Service permit would not violate water quality standards. The Forest Service argued that, under the CWA, only discharges from a point source or nonpoint source with a conveyance (i.e., a pipe or channel outlet) are regulated and, while cattle grazing may cause water pollution, it is not a regulated discharge under the act. However, in its ruling, the district court distinguished the definition of "discharge" from "discharge of a pollutant" from a point source and said that "pollution caused by cattle

[5] S. D. Warren Co. v. Maine Board of Environmental Protection, 126 S. Ct. 1843 (2006). For additional information, see CRS Report RS22429, *The State Role in the Federal Licensing of Hydropower Dams: S.D. Warren Co. v. Maine Board of Environmental Protection*, by Robert Meltz and Claudia Copeland.

[6] Oregon Natural Desert Association v. Thomas, 940 F.Supp. 1534, D.Or. 1996.

grazing constitutes a discharge into navigable waters within the meaning of section 401 of the Clean Water Act. Therefore, state certification under section 401 was required before the U.S. Forest Service issued a cattle grazing permit."

Supporters said that the ruling gave states new regulatory power over federal licenses or permits that affect water quality by clarifying that Section 401 applies to nonpoint source discharges of water pollution, in addition to point source discharges. Point sources are discrete conveyances, such as pipes or ditches, from which pollutants are discharged. Nonpoint source pollution is rainfall and snowmelt runoff from farmlands, ranches, city streets, and similar areas. The ruling had the potential to give states a stronger hand in determining how federal lands should be managed. If so, the impact on states could be significant, since cattle grazing is a common activity on millions of acres of western lands managed by the Forest Service and the U.S. Bureau of Land Management, and states could face a substantial workload in processing Section 401 certifications for hundreds of grazing permits annually. Additional impacts could occur if Section 401 were held to apply to other types of federally permitted activities generally categorized as nonpoint sources, such as timber harvesting or logging.

Federal agencies disagreed over how to respond to the Oregon district court's ruling. EPA favored letting the decision stand, on the basis that nonpoint source pollution is the most significant contributor to water pollution in many states, and the decision would give states more power to manage it. The Agriculture Department (parent of the Forest Service), on the other hand, urged the Department of Justice to support an industry group's appeal of the case, and ultimately the government did join in appealing the decision.

In 1998, a federal court of appeals reversed the district court's ruling, finding that cattle grazing on federal lands does not fall within the type of pollution covered by Section 401.[7] The court found that Congress intended to permit direct federal regulation of effluent flowing from point

[7] Oregon Natural Desert Association v. Dombeck, 151 F.3d 945 (CA9 1998).

sources, but to regulate nonpoint source pollution only through federal grants, not through Section 401 water quality certification. In November 1999, the Supreme Court declined to review the case.

The state of Oregon had responded to the 1996 district court decision by adopting rules establishing a certification process for livestock grazing permits on federal lands in Oregon. However, after the court of appeals reversed that ruling and the Supreme Court declined to review it, the state withdrew the rules. Groups representing ranchers, farmers, and similar interests were pleased that the district court's ruling was overturned, believing that Congress did not intend Section 401 to apply to nonpoint source pollution. Other CWA programs and tools such as financial incentives are better means of addressing nonpoint source pollution problems, some say. Environmentalists disagreed with the appeals court's conclusion and the legal outcome of the case, and many continue to argue that Section 401 generally supports a reading that includes nonpoint source discharges.

In a broader context, some observers had viewed the district court's ruling as giving a boost to ongoing activities in states to develop total maximum daily load (TMDL) allocations on pollution-impaired water bodies. Efforts to carry out this Clean Water Act requirement have been prompted by lawsuits in more than three dozen states, claiming that EPA and states have failed to fulfill mandates in the law.[8] In many cases, TMDLs are being developed that result in imposition of pollution control requirements and other measures affecting nonpoint sources as well as point sources in order to improve water quality and attain water quality standards. While the 9th Circuit's 1998 ruling did not directly affect the TMDL process, some persons saw the reversal of the lower court's ruling as removing one possible argument for ensuring that nonpoint sources are addressed in TMDL plans and processes.

Following the Supreme Court's 2006 ruling in the *S.D. Warren* case (discussed above), environmental advocates renewed legal challenge of grazing permits, arguing that the Court's ruling in that case that 401

---

[8] For additional information, see CRS Report R42752, *Clean Water Act and Pollutant Total Maximum Daily Loads (TMDLs)*, by Claudia Copeland.

certification applies to dam discharges could also be applied to agricultural runoff. The environmental group had challenged the failure of the Forest Service to require a company that sought a cattle grazing permit on Forest Service land to obtain a 401 certificate from the state of Oregon. However, the same federal court rejected the challenge, concluding that the facts in the more recent Oregon case differed from those in the *S.D. Warren* case, thus supporting its 1998 ruling that the CWA only allows states to address point source discharges under Section 401.[9]

LEGISLATIVE RESPONSE

Since the mid-1990s, Congress has shown interest in these issues in proposals reflecting varying perspectives, but no legislation that would modify Section 401 has been enacted.

In the 103rd Congress, interest in clarifying the scope of Section 401 certification authority led to several proposals. The Senate Environment and Public Works Committee included one such provision in S. 2093, a CWA reauthorization bill. It would have strengthened Section 401 by clarifying that applicants for a federal license or permit, including applicants for a FERC license to operate hydropower facilities, must obtain state certification that the project will comply with water quality standards and will allow for attainment and maintenance of designated uses included in the state's standards. The Senate did not act on S. 2093.

Following the Supreme Court's 1994 PUD No. 1 decision, disputes over Section 401 became an issue in the Congress. At the end of the 103rd Congress, legislation was introduced to amend the Clean Water Act and overturn the decision (S. 2566). The sponsor of the bill, Senator Wallop, said that the Court's decision threatened state water law (by limiting the amount of water that could be used for the project in question and, thus, interfering with state water rights systems) and the integrity of the FERC hydropower licensing process. The Senate did not act on this bill.

[9] Oregon Natural Desert Association v. U.S. Forest Service, 550 F.3d 778 (CA9 2008).

The 104th Congress addressed the issue in H.R. 961, a CWA reauthorization bill passed by the House in 1995. Section 507 would have made Section 401 inapplicable to hydropower projects if FERC were to determine that the state's certification is inconsistent with the Federal Power Act. The bill also would have set up a mechanism, to be administered by FERC, to resolve differences arising between the state and FERC on questions relating to the consistency of the 401 certification to a hydropower project. That is, in the event of a dispute between FERC and a state over 401 certification of a hydropower project, FERC would be authorized to resolve the dispute between itself and the state. This provision in H.R. 961 was one of several proposed to address the issue. Some Members favored simply exempting hydropower projects from Clean Water Act regulation, arguing that FERC project review is intended to consider inputs of state and federal agencies, Indian tribes, and the public in connection with licensing and relicensing decisions. Others argued that states should continue to have authority to regulate matters related to water quality concerns. Section 507 attempted to balance those concerns. However, no further action occurred on H.R. 961 during the 104th Congress, and similar legislation has not been proposed subsequently.

Legislative interest in Section 401 also occurred in connection with recommendations on national energy policy by Vice President Cheney's National Energy Policy Development Group in 2001. It recommended that the hydropower licensing process administered by FERC undergo administrative and legislative reform so that hydropower can contribute to meeting the nation's energy needs. At the same time, a 2001 FERC report concluded that the most common cause of delayed hydropower licensing proceedings is untimely receipt of state water quality certification under the Clean Water Act.[10] Responding to these concerns, legislation was proposed that would give applicants for hydropower licenses increased flexibility in complying with conditions imposed by federal agencies such

[10] Federal Energy Regulatory Commission, Staff Report to Congress, Hydroelectric Licensing Policies, Procedures, and Regulations, Comprehensive Review and Recommendations, May 2001, 145 pp.

as the Department of the Interior concerning, for example, the need for passageways through which fish can travel around a dam—another issue raised in the FERC report. The 109th Congress enacted the Energy Policy Act of 2005 (P.L. 109-58) with a provision (Section 241) requiring federal agencies to consider alternative license conditions proposed by the license applicant. While this provision only addressed the roles of federal agencies in hydropower licensing, not state certification under CWA Section 401, these water quality issues remain of interest to some stakeholders.

In the 112th Congress, the House passed legislation that would in part modify Section 401. The bill, H.R. 2018, the Clean Water Cooperative Federalism Act of 2011, would have amended several CWA provisions to restrict EPA's authority to provide oversight of states' implementation of aspects of the statute. The legislation was seen as a response to concerns that EPA has overreached in its statutory oversight of state actions on permitting and standard setting. Section 2(b) of the bill would have modified Section 401 to prohibit EPA from taking "any action to supersede the determination" of a state that a particular discharge will comply with applicable provisions of the CWA, including water quality standards. While under current law EPA has no explicit role or conditioning authority over a state's 401 certification, this provision of H.R. 2018 was apparently intended to deter EPA from overriding state determinations. In comments on the legislation, EPA officials questioned the intent of the provision, saying that the agency has not taken formal action to supersede a state's certification, making the practical effect of the provision unclear, in EPA's view. Bills with provisions similar to Section 2(b) of H.R. 2018 also were introduced in the 113th Congress (H.R. 1829 and H.R. 1948).

In the 114th Congress, Senate and House committees are considering a number of energy policy proposals, including legislation that some states believe could impinge on states' authority under CWA Section 401 concerning hydropower projects. The proposals include provisions intended to streamline review and approval of hydropower licensing projects. One bill, S. 1236, would designate FERC as the lead agency for coordinating all authorizations required under federal law—including 401 certification or other approvals required by federal and state agencies under federal law—

and would authorize FERC to establish schedules for issuance of all such authorizations. Under this bill, federal or state agencies responsible for such authorizations would be required to report annually to congressional committees on any of the agency's activities affecting a FERC proceeding, including a demonstration that the agency's actions meet policies established in the Federal Power Act. The federal or state resource agency would be required to maintain a publicly available website that tracks all information required in the annual report.

A second proposal, in a draft bill discussed at a House Energy and Commerce subcommittee hearing on May 13,[11] similarly would designate FERC as the lead agency for coordinating all federal and state authorizations required under federal law and would require cooperating agencies to comply with deadlines established by FERC. Further, the House draft bill would require that judicial review of federal and state agency actions related to hydropower projects, such as an order or action related to a 401 or other federal authorization, be heard exclusively in federal courts of appeal. Under current law, judicial challenges of 401 certifications are reviewed in state courts, not federal courts.

Several state agencies have raised concerns about aspects of these proposals. For example, Oregon officials said in a letter to House Energy and Commerce Committee leader that they believe that the draft House hydropower bill would "effectively strip the state's authority to protect its fish, wildlife, and water resources." Separately, Washington state officials said in a letter to House Commerce Committee leaders that they oppose the draft bill because "states' and tribes' authority to issue CWA 401 water quality certifications would be impacted or revoked." Similarly, in a letter to Senate Energy and Natural Resources Committee leaders, Maryland officials noted objections to S. 1236, saying, "In removing or impairing the states' primary role and responsibility under Section 401 to fashion conditions in FERC licenses, S. 1236 relegate[s] the states—the entities with the greatest interest and expertise in protecting water quality—to

---

[11] See http://docs.house.gov/meetings/IF/IF03/20150513/103443/BILLS-114pihDiscussion DraftonHydropowerRegulatoryModernization.pdf.

bystander or second-class status."[12] Press reports indicate that House and Senate committee leaders believe that CWA authority would not be affected by the bills in question.[13]

---

[12] Letters on file with the author of this report.

[13] Amena H. Saiyid, "Aides Say Hydropower Bills Will Streamline Permitting, Not Remove State Authority," *Daily Environment Report*, June 24, 2015, pp. A-10.

In: Water Quality
Editors: Edward Saunders et al.
ISBN: 978-1-53614-809-1

Chapter 3

EVOLUTION OF THE MEANING OF "WATERS OF THE UNITED STATES" IN THE CLEAN WATER ACT*

Stephen P. Mulligan

SUMMARY

The scope of waters that are properly the subject of federal water pollution legislation has been the subject of long-standing consideration by all three branches of the federal government, particularly in the aftermath of the 1972 amendments to the Federal Water Pollution Control Act, commonly referred to as the Clean Water Act. In a shift from early water pollution legislation, those amendments eliminated the requirement that the federally regulated waters—known as jurisdictional waters—must be navigable in the traditional sense, meaning that they are capable of being used by vessels in interstate commerce. Rather than use classical tests of navigability, the amendments redefined "navigable waters" for purposes of the Clean Water Act's jurisdiction to include "the waters of

* This is an edited, reformatted and augmented version of Congressional Research Service, Publication No. R44585, dated August 8, 2016.

the United States, including the territorial seas." Disputes over the proper meaning of that phrase have been ongoing.

Some courts and commentators also disagree on how the scope of federal jurisdictional waters changed over time as a result of interpretative approaches taken by the federal agencies responsible for administering the Clean Water Act—the Environmental Protection Agency (EPA) and the U.S. Army Corps of Engineers (Corps). While some believe EPA and the Corps consistently expanded the meaning of "waters of the United States," others contend that, in recent years, the agencies have construed the term in a narrower fashion than permitted under the Clean Water Act. In 2015, the Corps and EPA issued a new rule, known as the Clean Water Rule that substantially redefined "waters of the United States" in the agencies' regulations for the first time in more than two decades. Some observers disagree on whether the Clean Water Rule constitutes an expansion of jurisdiction over waters not previously regulated. This report provides context for this debate by examining the history of major changes to the meaning of "waters of the United States" as expressed in federal regulations, legislation, agency guidance, and case law.

The Clean Water Act uses the phrase "waters of the United States," but it does not include a statutory definition of that term. The long-standing disagreement over the meaning of that phrase has centered on the degree to which the Clean Water Act should be interpreted as covering the widest amount of "waters" that could permissibly be federally regulated under the Constitution, or whether that term should be interpreted in a more limited fashion.

Federal authority to regulate waters within the United States primarily derives from the Commerce Clause, and accordingly, federal laws and regulations concerning waters of the United States cannot cover matters which exceed that constitutional source of authority. During the first two decades after the passage of the Clean Water Act, courts generally interpreted the act as having a wide jurisdictional reach. In recent decades, however, the Supreme Court has emphasized that "the grant of authority to Congress under the Commerce Clause, though broad, is not unlimited." This modern Commerce Clause jurisprudence has informed federal courts' approach to interpreting which "waters" are subject to the Clean Water Act.

Most recently, courts have taken up legal challenges to the Clean Water Rule. On October 9, 2015, the United States Court of Appeals for the Sixth Circuit stayed its enforcement, and the House version of the FY2017 Interior-Environment appropriations bill (H.R. 5538) would block its application by prohibiting the use of appropriated funds to implement changes to the meaning of jurisdictional waters beyond those that were in effect on October 1, 2012.

INTRODUCTION

The scope of waters that are properly the subject of federal water pollution legislation has been the subject of long-standing consideration by all three branches of the federal government, particularly in the aftermath of the 1972 amendments to the Federal Water Pollution Control Act, commonly referred to as the Clean Water Act. In a shift from prior water pollution legislation, those amendments eliminated the requirement that the federally regulated waters—known as jurisdictional waters—must be navigable in the traditional sense,[1] meaning that they are capable of being used by vessels in interstate commerce.[2] Rather than use classical tests of navigability developed in the 19th century,[3] the Clean Water Act redefined

---

1 In *Riverside Bayview Homes v. United States*, 474 U.S. 121, 132-33 (1985), the Supreme Court explained that, in the 1972 Clean Water Act amendments, "Congress evidently intended to repudiate limits that had been placed on federal regulation by earlier water pollution control statutes, and to exercise its powers under the Commerce Clause to regulate at least some waters that would not be deemed 'navigable' under the classical understanding of that term." The Commerce Clause gives Congress the power to "regulate commerce with foreign nations, and among the several states...." U.S. CONST. art. 1, §8, cl. 3.

2 *See The Daniel Ball*, 77 U.S. 557, 563 (1870) (construing the term "navigable waters," as employed in federal statutes at issue, as covering those waters that are "used or are susceptible of being used in their ordinary condition as highways for commerce over which trade and travel are or may be conducted in the customary modes of trade and travel on water"); The Montello, 87 U.S. 430, 441-42 (1874) ("If [the subject water] be capable in its natural state of being used for purposes of commerce, no matter in what mode the commerce may be conducted, it is navigable in fact, and becomes in law a public river or highway."); United States v. Holt State Bank, 270 U.S. 49 (1926) ("The rule long since approved by this court in applying the Constitution and laws of the United States is that streams or lakes which are navigable in fact must be regarded as navigable in law; that they are navigable in fact when they are used, or are susceptible of being used, in their natural and ordinary condition, as highways for commerce, over which trade and travel are or may be conducted in the customary modes of trade and travel on water[.]").

3 At common law, only waters subject to the ebb and flow of tide were held to be navigable waters subject to federal jurisdiction. *See The Daniel Ball*, 77 U.S. at 563; Nelson v. Leland, 63 U.S. 48, 55 (1860). This rule was largely due to the fact that, given the geography of England, there were few waters which were susceptible to use in commerce that were not also subject to the ebb and flow and tide. *See The Daniel Ball*, 77 U.S. at 563. Based on geographic differences and the recognition that "[s]ome of our [American] rivers are as navigable for many hundreds of miles above as they are below the limits of tide water," American courts in the 19th century departed from the common law rule and began to analyze whether waters were "navigable-in-fact." *Id.*; *see also, e.g.*, *Nelson*, 63 U.S. at 55-56 (distinguishing between admiralty jurisdiction exercised in England and in the United States); Escanaba Cnty. v. Chicago, 107 U.S. 678, 682-83 (1883) (describing how the common law rule "has long since been discarded in this country").

"navigable waters" for purposes of federal regulatory jurisdiction to include "the waters of the United States, including the territorial seas."[4] Disputes over the proper meaning of that phrase have been ongoing.

Some courts and commentators also disagree on how the scope of federal jurisdictional waters changed over time as a result of interpretative approaches taken by the agencies responsible for administering the Clean Water Act—the Environmental Protection Agency (EPA) and the U.S Army Corps of Engineers (Corps). While some believe EPA and the Corps consistently expanded the meaning of "waters of the United States," others contend that, in recent years, the agencies have construed the term in a narrower fashion than permitted under the Clean Water Act.[5] This debate resurfaced most recently in May 2015 when the Corps and EPA issued a rule, known as the Clean Water Rule, which substantially redefined "waters of the United States" in the agencies' regulations for the first time in more than two decades.[6] EPA and the Corps contend that the Clean Water Rule governs only waters that have been historically regulated under the Clean Water Act, but its opponents argue that it constitutes an unlawful expansion of authority beyond that which is allowed in the act or the Constitution.[7] This report provides context for this debate by examining the history of major changes to the meaning of "waters of the United States" as

---

[4] *See* 33 U.S.C. §1362(7).

[5] *Compare, e.g.*, Rapanos v. United States, 547 U.S. 715, 722 (2006) (plurality opinion) (Scalia, J.) (describing "the immense expansion of federal regulation of land use that has occurred under the Clean Water Act—without any change in the governing statute—during the past five Presidential administrations") *and* Jamison E. Colburn, *Waters of the United States, Theory, Practice, and Integrity at the Supreme Court*, 34 FLA. ST. U. L. REV. 183, 199 (2007) (explaining "how two relatively conservative administrative agencies gradually decided, in six different Presidential administrations, to expand federal jurisdiction as dramatically as they have") *with* Jon Devine et al., *The Historical Scope of Clean Water Act Jurisdiction*, ENVTL. FORUM, July/August 2012, at 57, (attempting to "refute[] the contention that the Corps and EPA have steadily expanded their assertions of the [Clean Water] [A]ct's scope" and arguing "that the agencies have actually retreated from the jurisdictional scope initially intended and asserted for the CWA") *and* Envtl. Prot. Agency & Dep't of the Army, Technical Support Document for the Clean Water Rule: Definition of Waters of the United States, 18-34 (May 27, 2015), https://www.epa.gov/sites/production/files/2015-05/documents/technical_support_document_for_the_clean_water_rule_1.pdf [hereinafter Technical Support for the Clean Water Rule] (discussing the history of regulatory definitions of "waters of the United States" and asserting that the Clean Water Rule represents a contraction in jurisdiction).

[6] *See infra* "The Clean Water Rule."

[7] *See infra* "Response to the Clean Water Rule."

expressed in federal regulations, legislation, agency guidance, and case law.[8]

BACKGROUND

The Clean Water Act is the principal law governing pollution of the nation's surface waters.[9] Among other requirements, the act prohibits the unauthorized discharge of pollutants into "navigable waters,"[10] and requires persons wishing to discharge dredged or fill material into "navigable waters" to obtain a permit from the Corps.[11] In its definition

---

[8] While this report outlines many notable changes to the definition of "waters of the United States," it does not address every agency interpretation or application of that phrase. For example, the report does not address most property-specific applications of the definition of "waters of the United States," such as those made in National Pollutant Discharge Elimination System Permit (NPDES) decisions, e.g., Envtl. Prot. Agency, Off. of Gen. Counsel, Opinion No. 21, In re Riverside Irrigation District, LTD and 17 Others (June 27, 1975), 1975 WL 23864, at *1-5 (interpreting the Clean Water Act and EPA's definition of "waters of the United States" in its regulations to determine whether an NPDES permit may be required for irrigation return flow canals and irrigation and drainage ditches); Section 404 dredge and fill permit decisions, ORM Jurisdictional Determinations and Permit Decisions, U.S. ARMY CORPS OF ENG'RS, http://corpsmapu.usace.army.mil/cm_apex/f?p=340:2:0::NO (last visited July 28, 2016) (database of Section 404 permit decisions), or jurisdictional determinations, ORM Jurisdictional Determinations and Permit Decisions, U.S. Army Corps of Eng'rs, http://corpsmapu.usace.army.mil/cm_apex/f?p=340:11:0::NO (last visited July 28, 2016) (database of jurisdictional determinations); U.S. Army Corps of Eng'rs v. Hawkes Co., Inc., 578 U.S. ___, 136 S. Ct. 1807 (2016) (providing background on the process of providing jurisdictional determinations as to whether a specific parcel is subject to Section 404's dredge and fill permit requirements). This report also does not address minor changes among EPA's various regulatory definitions which do not reflect a change in the agency's overall interpretation of the scope of waters of the United States. *See, e.g.*, Envtl. Prot. Agency, Off. of Gen. Counsel, Opinion No. 77-3, Clarification of the Term "Navigable Waters" as it is Presently Used in FWPCA Regulations and Guidelines (February 28, 1977), 1977 WL 28236 (discussing differences among EPA's definitions and proposals to streamline the definition).

[9] *See* Michael Goldman, Symposium, *Drilling into Hydraulic Fracturing and Shale Gas Development: A Texas and Federal Environmental Perspective*, 19 TEX. Wesleyan L. Rev. 185, 191 (2012); Walter G. Wright, Jr. & Albert J. Thomas III, *The Federal/Arkansas Water Pollution Control Programs: Past Present and Future*, 23 U. Ark. Little rock L. Rev. 541, 550 (2001); CRS Report RL30030, *Clean Water Act: A Summary of the Law*, by Claudia Copeland, at 1.

[10] 33 U.S.C. §1311(a).

[11] *Id.* §1344.

section, the act defines the term "navigable waters" to mean "waters of the United States, including its territorial seas."[12] This single, jurisdiction-defining phrase applies to the entire law, including the national pollutant discharge elimination system (NPDES) permit program;[13] permit requirements for disposal of dredged or fill material, known as the Section 404 program;[14] water quality standards and measures to attain them;[15] oil spill liability and prevention;[16] and enforcement.[17]

Key Terminology

Navigable-in-fact waters: A term of art developed by courts to describe waters that are navigable in the traditional sense, meaning they are capable of being used by vessels in interstate commerce.[18]

Navigable waters of the United States: A statutory phrase often used in pre-Clean Water Act legislation, which courts generally interpreted to mean navigable-infact waters.[19]

Interstate waters: Waters which form a part of state's boundary.[20]

Navigable waters: An anomalous term as used in the Clean Water Act. The Clean Water Act governs "navigable waters,"[21] but this phrase is defined within the statute such that it is not limited to waters that are navigable-in-fact.

Waters of the United States: The jurisdiction-defining phrase in the Clean Water Act. That statute generally regulates "navigable waters," but it defines that term to mean "the Waters of the United States, including the territorial seas."[22]

Jurisdictional waters: A term of art used by courts to describe those waters subject to federal regulatory jurisdiction under the Clean Water Act.[23]

12 *Id.* §1362(7).
13 *Id.* §1342.
14 *Id.* §1344.
15 *Id.* §1313.
16 *Id.* §1321.
17 *Id.* §1319.
18 *See, e.g.*, PPL Mont., LLC v. Montana, 565 U.S. __, 132 S. Ct. 1215, 1219 (2012) (quoting *The Daniel Ball*, 77 U.S. 557, 563 (1871)).
19 *See, e.g., The Daniel Ball*, 77 U.S. at 563.
20 *See* Federal Water Pollution Control Act of 1948, P.L. 845, 62 Stat. 1155, 1161.
21 33 U.S.C. §1362(7).
22 *Id.*
23 *See, e.g.*, U.S. Army Corps of Eng'rs v. Hawkes Co., Inc., 578 U.S. ___, 136 S. Ct. 1807, 1814 (2016).

The Clean Water Act itself does not expand further on the meaning of "waters of the United States." Instead, the Corps and EPA have expounded on this phrase through agency guidance and regulations, which, at various times, have been stricken down or modified as a result of legal challenges. These legal challenges—particularly those which were successful—can be seen to have followed broader trends in interpretation of the Commerce Clause,[24] which gives Congress the power to "regulate commerce with foreign nations, and among the several states...."[25]

Federal court review of the Corps' and EPA's interpretation of which "waters" are subject to the Clean Water Act has been informed by both statutory and constitutional considerations, including whether an agency's interpretation could potentially enable it to regulate matters beyond the constitutional reach of the federal government as interpreted in recent Supreme Court case law. Federal authority to regulate waters within the United States primarily derives from the Commerce Clause.[26] Accordingly, federal laws and regulations concerning waters of the United States cannot cover matters which exceed that constitutional source of authority.[27] For a period after its enactment in 1972, courts generally interpreted the Clean Water Act as having a wide jurisdictional reach, but, in recent decades, the Supreme Court has emphasized that "the grant of authority to Congress under the Commerce Clause, though broad, is not unlimited."[28]

A time line of events in the evolution of the definition of "waters of the United States" is provided in the *Appendix*, and major events are shown in *Figure 1.*

---

[24] *See infra* "Judicially Imposed Limitations Beginning in the Late 1990s."

[25] U.S. CONST. art. 1, §8, cl. 3.

[26] *See* Gilman v. Philadelphia., 70 U.S. 713, 724-725 (1866) ("The power to regulate commerce comprehends the control for that purpose, and to the extent necessary, of all the navigable waters of the United States which are accessible from a State other than those in which they lie."); Gibbons v. Ogden, 22 U.S. 1 (1824) (concluding that Congress had authority under the Commerce Clause to license steamboat operations in New York waters); see also Riverside Bayview Homes, Inc., 474 U.S. 121, 133 (1985) (discussing Congress's invocation of the Commerce Clause powers in enacting the Clean Water Act).

[27] *See, e.g.*, Solid Waste Agency of N. Cook Cnty. v. U.S. Army Corps of Eng'rs, 531 U.S. 159, 173-74 (2001) (declining to interpret jurisdictional reach of the Clean Water Act in a manner which may exceed the limits of the Commerce Clause).

[28] *See id.* at 172 (citing United States v. Morrison, 529 U.S. 598 (2000); United States v. Lopez, 514 U.S. 549 (1995)).

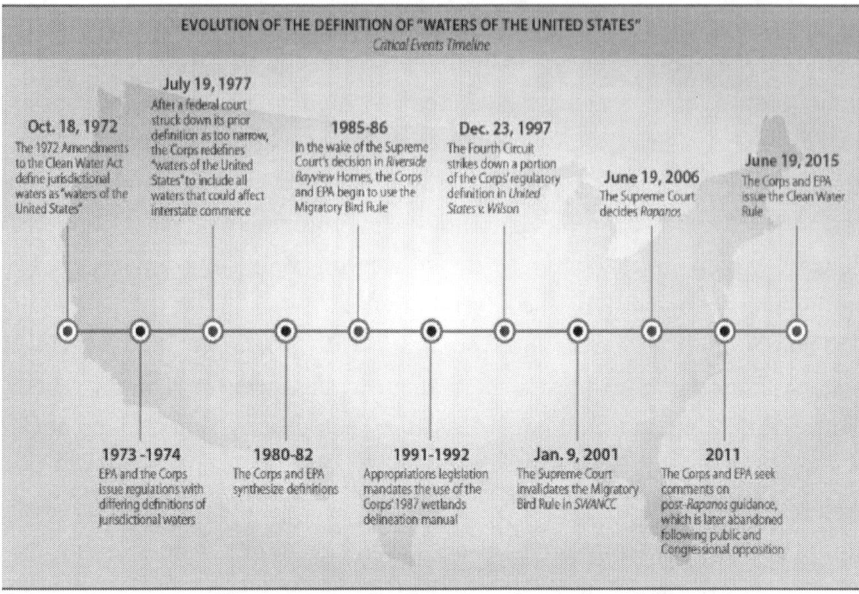

Source: Congressional Research Service, based on the sources cited in this report.

Figure 1. Major Events in the Evolution of "Waters of the United States."

THE EARLY HISTORY OF JURISDICTIONAL WATERS

Historically, federal laws regulating waterways, such as the Rivers and Harbors Appropriations Act of 1899 (Rivers and Harbors Act), exercised jurisdiction over "navigable water[s] of the United States[.]"[29] The Supreme Court interpreted this phrase to govern only waters that were "navigable-in-fact"—meaning that they were "used, or are susceptible of being used, ... as highways for commerce, over which trade and travel are

[29] *See* Rivers and Harbors Appropriations Act of 1899, 30 Stat. 1121, 1151 (codified in 33 U.S.C. §401) ("It shall not be lawful to construct or commence the construction of any bridge, causeway, dam, or dike over or in any ... navigable water of the United States until the consent of Congress shall have been obtained...."); An Act to Provide Security for the Lives of Passengers on Board of Vessels Propelled by Steam, 5 Stat. 304 (1838) (providing that "it shall not be lawful for the owner ... of any steamboat ... to transport any goods, wares, merchandise or passengers, in or upon ... navigable waters of the United States ... without having first obtained ... a license").

or may be conducted in the customary modes of trade and travel on water."[30]

Beginning with the Federal Water Pollution Control Act of 1948, Congress began to use a different jurisdiction-defining phrase in which it regulated "interstate waters," defined as "all rivers, lakes, and other waters that flow across, or form a part of, a State's boundaries."[31] That legislation was amended in 1961 to expand federal jurisdiction from "interstate waters" to "interstate or navigable waters[.]"[32]

The Federal Water Pollution Control Act Amendments of 1972,[33] which have commonly become known as the Clean Water Act,[34] also amended the jurisdictional reach of federal water pollution legislation. There, Congress again exercised jurisdiction over "navigable waters," but provided a new definition of that phrase that was not used in prior legislation, stating the following: "The term 'navigable waters' means the waters of the United States, including the territorial seas."[35] This subtle definitional change proved to have tremendous consequences for the jurisdictional scope of the Clean Water Act.

In debating the 1972 amendments that created the Clean Water Act, some Members of Congress explained that the revised definition was intended to expand the law's jurisdiction beyond traditionally navigable or interstate waters. The conference report states that the "conferees fully intend that the term 'navigable waters' be given the broadest possible

---

[30] *See The Daniel Ball*, 77 U.S. 557, 563 (1870). Waters, the Court explained in *The Daniel Ball*, "constitute navigable waters of the United States within the meaning of the acts of Congress, in contradistinction from the navigable waters of the States, when they form in their ordinary condition by themselves, or by uniting with other waters, a containued [sic] highway over which commerce is or may be carried on with other States or foreign countries in the customary modes in which such commerce is conducted by water." *Id.*

[31] *See* Federal Water Pollution Control Act of 1948, P.L. 845, §10(e), 62 Stat. 1155, 1161.

[32] *See* Federal Water Pollution Control Act Amendments of 1961, P.L. 87-88, §8(a), 75 Stat. 204, 208 (codified in 33 U.S.C. §1160(a) (1970)).

[33] P.L. 92-500, 86 Stat. 816.

[34] *See* History of the Clean Water Act, ENVTL. PROT. AGENCY (May 25, 2016), https://www.epa.gov/laws-regulations/history-clean-water-act.

[35] *See* Federal Water Pollution Control Act Amendments of 1972, P.L. 92-500, §502(7), 86 Stat. 816, 886 (codified in §1362(7)).

constitutional interpretation....”[36] And during debate in the House on approving the conference report, one Representative explained that the definition “clearly encompasses all water bodies, including streams and their tributaries, for water quality purposes.”[37] Courts have frequently referred to this legislative history when interpreting the scope of the Clean Water Act.[38]

DIFFERING AGENCY DEFINITIONS FOLLOWING THE CLEAN WATER ACT

The Corps and EPA share responsibilities for administering the Clean Water Act. Both agencies have administrative responsibilities under Section 404 of the act,[39] and EPA exclusively administers most other Clean Water Act-related programs.[40] As a consequence of this dual jurisdiction, both agencies create regulations defining the waters subject to their

---

[36] S.Rept. 92-1236, at 144 (1972) (Conf. Rep.).

[37] *See* 118 Cong. Rec. 33,757 (1972) (statement of Rep. Dingell).

[38] *See, e.g.*, Solid Waste Agency of N. Cook Cnty. v. U.S. Army Corps of Eng'rs, 531 U.S. 159, 180-81 (2001) (Stevens, J., dissenting) (discussing the conference report and changes to the House version of the 1972 amendments); Riverside Bayview Homes v. United States, 474 U.S. 121, 132-33 (1985) (citing, among other things, the conference report and statements of Rep. Dingell); United States v. Holland, 373 F. Supp. 665, 667-73 (M.D. Fla. 1974) (discussing legislative history of the 1972 amendments).

[39] *See* Advance Notice of Proposed Rulemaking on the Clean Water Act Regulatory Definition of “Waters of the United States,” 68 Fed. Reg. 1991, 1991-92 (January 15, 2003) (providing background on the agencies' roles in administering the Clean Water Act); Bradford C. Mank, *The Murky Future of the Clean Water Act after SWANCC: Using a Hydrological Connection Approach to Saving the Clean Water Act*, 30 ECOLOGY L.Q. 811, 814 n.6 (2003) (discussing the division of administrative responsibility under the Clean Water Act); CRS Report RL30030, *supra* note 9, at 6 (discussing the joint administration of Section 404).

[40] *See* 33 U.S.C. §1251(d) (stating that EPA will implement the Clean Water Act unless expressly stated otherwise); Benjamin R. Civiletti, *Administrative Authority to Construe § 404 of the Federal Water Pollution Control Act*, 43 Op. Att'y Gen. 197, 197 (1979) (“Congress intended to confer upon the administrator of the [EPA] the final administrative authority to determine ... the reach of the term ‘navigable waters’....”).

regulatory jurisdiction.[41] In the initial years following the enactment of the Clean Water Act, their respective definitions differed significantly.[42]

EPA's Initial Definition

EPA issued its first internal definition of jurisdictional waters in a February 6, 1973, memorandum from its Office of the General Counsel.[43]

EPA's First Internal Definition of Jurisdictional Waters[44]

In a February 6, 1973, memorandum from its Office of the General Counsel, EPA proposed to define jurisdictional waters through six categories:

1. all navigable waters of the United States;
2. tributaries of navigable waters of the United States;
3. interstate waters;
4. interstate lakes, rivers, and streams which are utilized by interstate travelers for recreational or other purposes;
5. interstate lakes, rivers, and streams from which fish or shellfish are taken and sold in interstate commerce; and
6. interstate lakes, rivers, and streams which are utilized for industrial purposes by industries in interstate commerce.

---

[41] *See* 33 C.F.R. §328.3 (2016) (containing the Corps' definition of "waters of the United States"); 40 C.F.R. §122.2 (2016) (including one EPA definition of "waters of the United States").

[42] *See* Bradford Mank, *Implementing Rapanos—Will Justice Kennedy's Significant Nexus Text Provide a Workable Standard for Lower Courts, Regulators and Developers?*, 40 IND. L. REV. 291, 300 (2007) ("From 1972 until 1975, the EPA and the Corps disagreed about the scope of the [Clean Water] Act's jurisdiction.").

[43] *See* Envtl. Prot. Agency, Off. of Gen. Counsel, Meaning of the Term "Navigable Waters" (February 13, 1973), 1973 WL 21937. Prior to the General Counsel's memorandum, EPA published a notice of proposed rule in which it repeated, without expanding upon, the statutory definition of "navigable waters." *See* Notice of Proposed Rulemaking, National Pollutant Discharge Elimination System, 38 Fed. Reg. 1362, 1363 (January 11, 1973). Both the Corps and EPA initially addressed the definition of "navigable waters" in their Clean Water Act regulations rather than elaborate on the meaning of "waters of the United States."

[44] *See* Envtl. Prot. Agency, Off. of Gen. Counsel, Meaning of the Term "Navigable Waters" (February 13, 1973), 1973 WL 21937.

EPA, however, never incorporated this definition in its regulations. Instead, it published its first set of regulations implementing the Clean Water Act's NPDES program later in 1973, in which it largely adopted the general counsel's recommended definition, but with one critical change: EPA revised categories four through six to include *intrastate* lakes, rivers, and streams which are utilized for interstate activities.[45]

The Corps' Initial Definition

The Corps' early implementation of the 1972 amendments differed considerably from EPA's regulations.[46] After initially proposing regulations that simply repeated the statutory definition of "navigable waters,"[47] the Corps issued final regulations in April 1974.[48] There, the Corps acknowledged the language from the conference report for the Clean Water Act as calling for the "broadest possible constitutional interpretation" of navigable waters, but concluded that the Constitution limited its jurisdiction to the same waters which it regulated under preexisting laws, such as the Rivers and Harbors Act.[49] Based on this

---

[45] *See* National Pollutant Discharge Elimination System, 38 Fed. Reg. 13528, 13,529 (1973) (codified at 40 C.F.R. §125.1(p) (1974)). EPA issued a similar, but slightly modified definition of "navigable waters" in its regulations implementing the 1972 amendments' oil pollution prevention provisions. *See* Oil Pollution Prevention, 38 Fed. Reg. 34,164, 34,165 (December 11, 1973) (codified in 33 C.F.R. §112.2(k) (1974)). This definition did not include intrastate waters used for industrial purposes interstate commerce (Category 6), and it expanded upon on the first category as follows: "all navigable waters of the United States, as defined in judicial decisions prior to passage of the 1972 amendments...." *see id.* Other EPA regulations at the time repeated the statutory definition of "navigable waters" without expanding upon it. *See* Envtl. Prot. Agency, Off. of Gen. Counsel, Opinion No. 77-3, Clarification of the Term "Navigable Waters" as it is Presently Used in FWPCA Regulations and Guidelines (February 28, 1977), 1977 WL 28236 (discussing differences among EPA's definitions).

[46] *See* Mank, *supra* note 42, at 300.

[47] *See* Proposed Policy, Practice, and Procedure: Permits for Activities in Navigable Waters or Ocean Waters, 38 Fed. Reg. 12,217, 12,218 (May 10, 1973).

[48] Permits for Activities in Navigable Waters or Ocean Waters, 39 Fed. Reg. 12,115, 12,119 (April 3, 1974) (codified in 33 C.F.R. §209.120(d)(1) (1974)).

[49] *See id.* at 12115.

reasoning, the Corps defined "navigable waters" using language that generally limited its jurisdiction to waters that were navigable-infact.[50]

The Corps' First Definition of Jurisdictional Waters

In its first set of final regulations implementing Section 404, the Corps equated the "navigable waters" regulated under the Clean Water Act with traditionally navigable waterways regulated under preexisting federal laws like the Rivers and Harbors Act:

The term "navigable waters of the United States" and "navigable waters," as used herein mean those waters of the United States which are subject to the ebb and flow of the tide, and/or are presently, or have been in the past, or may be in the future susceptible for use for purposes of interstate or foreign commerce....[51]

CALLAWAY AND ITS AFTERMATH

Less than one year after the Corps published its first regulations defining jurisdictional waters, the United States District Court for the District of Columbia struck them down as too narrow and inconsistent with the Clean Water Act.[52] In Natural Resources Defense Council v. Callaway, the court held that because "Congress ... asserted federal jurisdiction over the nation's waters to the maximum extent permissible under the Commerce Clause of the Constitution[,]" the definition could not be limited to "traditional tests of navigability[.]"[53] The court ordered the

[50] *See* 33 C.F.R. §209.12(d)(1) (1974); *see also* 33 C.F.R. §209.260(e)(1) (1974) ("[I]t is the water body's capability of use by the public for purposes of transportation or commerce which is the determinative factor.").

[51] 33 C.F.R §209.12(d)(1) (1974).

[52] *See* Nat. Res. Def. Council, Inc. v. Callaway, 392 F. Supp. 685 (D.D.C. 1975).

[53] *See* Callaway, 392 F. Supp. 685 Courts prior to *Callaway* also concluded that regulatory jurisdiction under the 1972 Amendments extended to waters that were not navigable-in-fact. *See, e.g.*, United States v. Ashland Oil & Transp. Co., 504 F.2d 1317, 1129-30 (6th Cir. 1974) (holding that Congress intended to control discharge of pollutants into nonnavigable tributaries which flowed into navigable waters, and that this exercise of authority was constitutional); United States v. Holland, 373 F. Supp. 665, 673 (M.D. Fla. 1974) (holding that Congress intended to address "the pollution of non-navigable mosquito canals and mangrove wetland areas").

Corps to produce new regulations which acknowledged "the full regulatory mandate" of the Clean Water Act.[54]

The Corps' Expansion of Jurisdictional Waters Following *Callaway*

The Corps responded to *Callaway* on May 6, 1975, by publishing proposed regulations which offered four alternative methods of redefining the Corps' jurisdiction under the 1972 amendments.[55]

The Corps' Four Proposed Alternatives Following *Callaway*

Following *Callaway*, the Corps published four proposed alternative scenarios in which it would evaluate Section 404 permits:

Alternative 1: Extend the Corps' jurisdiction to "virtually every coastal and inland artificial or natural waterbody[,]" and apply the Corps' permitting process to "all disposal of dredged or fill material in virtually every wetland contiguous to coastal waters, rivers, estuaries, lakes, streams and artificial waters...."

Alternative 2: Limit jurisdiction to waters subject to the ebb and flow of tide and navigablein-fact inland waters and their primary tributaries.

Alternative 3: Apply the jurisdictional authority in Alternative 1, but utilize only the Corps' standard permitting process for navigable-in-fact waters. For waters that are not navigable-infact,[56] the Corps would approve permits unless the state objects.

Alternative 4: Apply the limited jurisdiction in Alternative 2 and the limited permitting process of Alternative 3. The Corps stated that Alternative 4 was its preferred approach.[57]

---

[54] *See* Callaway, 392 F. Supp. at 685. Although the court ordered the Corps to publish final regulations within 30 days, it later extended its original deadlines. *See* Proposed Policy, Practice and Procedure: Permits for Activities in Navigable Waters or Ocean Waters, 40 Fed. Reg. 19,766 (May 6, 1975).

[55] *See* Proposed Policy, Practice and Procedure: Permits for Activities in Navigable Waters or Ocean Waters, 40 Fed. Reg. at 19,766.

[56] In its early regulations implementing the Clean Water Act, the Corps did not use the phrase "navigable-in-fact," and instead used the phrase "navigable waters of the United States," which was derived from prior laws such as the Rivers and Harbors Act. *See id.* at 17,968. Because of the similarity of language, and to provide clarity, this report uses the phrase "navigable-in-fact" to refer to traditionally navigable waters.

At the same time that it proposed these alternatives, the Corps published a press release stating that the holding of *Callaway* may require "the rancher who wants to enlarge his stock pond, or the farmer who wants to deepen an irrigation ditch or plow a field, or the mountaineer who wants to protect his land against stream erosion" to obtain federal permits.[58] These events brought significant public and media attention to the breadth of jurisdiction under the Clean Water Act.[59] It also created a disagreement between the Corps and EPA,[60] and led to a series of subcommittee hearings in the House and Senate.[61]

In the aftermath of this attention, the Corps issued interim final regulations in 1975 in which it revised the definition of "navigable waters" by adopting much of the structure used in EPA's 1973 regulations.[62] The Corps' definition also added "wetlands, mudflats, swamps, marshes, and shallows" that are "contiguous or adjacent to other navigable waters" and

---

[57] *See id.* at 19,766.

[58] *See* Press Release, Dep't of the Army, Office of the Chief of Eng'rs (May 6, 1975), reprinted in *Section 404 of the Federal Water Pollution Control Act Amendments of 1976, Hearings Before the Senate Comm. on Public Works*, 94th Cong., 517 (1976).

[59] *See, e.g.*, *Army Engineers Seek Control of All Waters, Down to Ponds*, NEW YORK TIMES, May 7, 1975, at 12, PROQUEST; *Wetlands and the Corps of Engineers*, WASHINGTON POST, June 3, 1975, at A18, PROQUEST. The Corps received over 4,500 comments on its proposed regulations, including comments from a "large number of Governors; members of Congress; Federal, State, and local agencies;" interest groups and members of the public. *See* Permits for Activities in Navigable Waters or Ocean Waters, 40 Fed. Reg. 31,320 (July 25, 1975).

[60] EPA responded to the press released by accusing the Corps of misleading the public. Letter from Russell E. Train, EPA Admin., to Lt. Gen. William C. Gribble, Jr., Chief of Eng'rs, U.S. Army Corps. of Eng'rs (May 16, 1975), reprinted in *Section 404 of the Federal Water Pollution Control Act Amendments of 1972, Hearings Before the Senate Comm. on Public Works*, 94th Cong., 528-29 (1976) (stating that the public confusion and misunderstanding "is directly attributable to the seriously inaccurate and misleading press release issued by the Corps").

[61] *See Development of New Regulations by the Corps of Engineers, Implementing Section 404 of The Federal Water Pollution Control Act Concerning Permits for Disposal of Dredge or Fill Material: Hearings Before the Subcomm. on Water Resources of the H. Comm. On Public Works and Transportation*, 94th Cong. (1995); *see also* Joanne M. Frasca, *Federal Control of Wetlands: The Effectiveness of Corps' Regulations under 404 of the FWPCA*, 51 NOTRE DAME LAW. 505, 506 & n.11 (1976) (citing and discussing hearings held on July 15, 16, and 22, 1975).

[62] *Compare* Permits for Activities in Navigable Waters or Ocean Waters, 40 Fed. Reg. 31,320, 31,324 (July 25, 1975) [hereinafter 1975 Interim Final Rule] (codified in 33 C.F.R. §209.129(d)(2) (1976)) (Corps' revised definition) *with* 40 C.F.R. §125.1(p) (1974) (EPA's definition of "navigable waters" for purposes of the Clean Water Act oil pollution prevention program).

"artificially created channels and canals used for recreational or other navigational purposes that are connected to other navigable waters" to the definition of "waters of the United States."[63]

The Corps' 1977 Regulations

In 1977, the Corps issued final regulations reorganizing the definition of "waters of the United States" into five categories.[64]

The final category of the 1977 definition contained the Corps' most expansive definition of jurisdictional waters as of that time. A footnote to the Corps' regulations explained that the Category Five waters incorporate "all other waters of the United States that could be regulated under the Federal government's Constitutional powers to regulate and protect interstate commerce."[65] The Corps would continue to use this Commerce Clause-focused provision (with revisions) in its regulations through 2014,[66] and EPA would later adopt it in its regulations.[67]

---

[63] *See* 1975 Interim Final Rule, 40 Fed. Reg. at 31,324. The 1975 Interim Final Rule used a phased approach in which the Corps expanded its authority in three phases to be completed by 1977. See *id.* at 31,325-26. Phase I, which was immediately effective, included coastal waters and inland navigable-in-fact waters and their adjacent wetlands. *See id.* at 31, 321-26. Phase II, which took effect on July 1, 1976, extended to lakes and primary tributaries of Phase I waters, as well as wetlands adjacent to the lakes and primary tributaries. *Id.* Phase III, which took effect on July 1, 1977, extended to all remaining areas encompassed by the regulations. *Id.* at 31,325.

[64] *See* Final Rule, Regulatory Programs of the Corps of Engineers, 42 Fed. Reg. 37,122 (July 19, 1977) [hereinafter 1977 Corps Rule]. Rather than continue to adjust the meaning of "navigable waters," the Corps expanded upon the meaning of the phrase "waters of the United States" as its method of defining its regulatory jurisdiction in the 1977 Corps Rule. *See id.* at 37,127 ("Many suggested that we change our nomenclature of the term 'navigable waters' and refer to our jurisdiction under Section 404 [of the Clean Water Act] as 'waters of the United States.' ... We have adopted this suggestion and feel that it will assist in distinguishing between the Section 404 program and the types of waters that are subject to the permit programs administered under the [Rivers and Harbors Act].").

[65] *See* 33 C.F.R. §323.2(a)(5) n.2 (1978).

[66] *See* 33 C.F.R. §328.3(a)(3) (2014) (defining "waters of the United States" to include, among other things, "[a]ll other waters such as intrastate lakes, rivers, streams (including intermittent streams), mudflats, sandflats, wetlands, sloughs, prairie potholes, wet meadows, playa lakes, or natural ponds, the use, degradation or destruction of which could affect interstate or foreign commerce").

[67] *See* 40 C.F.R. §122.3 (1981).

The Corps' 1977 Definition and Its Commerce Clause-Focused Provision

The Corps reorganized the definition of "waters of the United States" in 1977, with Category 5 waters containing its broadest definition of jurisdictional waters as of that date:

1. The territorial seas with respect to the discharge of fill material ... ;
2. Coastal and inland waters, lakes, rivers, and streams that are [navigable-in-fact], including adjacent wetlands;
3. Tributaries to navigable waters of the United States ... ;
4. Interstate waters and their tributaries, including adjacent wetlands; and
5. All other waters of the United States not identified in Categories 1-3, such as isolated lakes and wetlands, intermittent streams, prairie potholes, and other waters ... the destruction of which could affect interstate commerce.[68]

THE CLEAN WATER ACT OF 1977

After the Corps' 1975 and 1977 regulations, bills were introduced which sought to limit the Clean Water Act's jurisdiction to traditional, navigable-in-fact waters.[69] This limiting legislation never became law. Instead, Congress amended the Federal Water Pollution Control Act through the Clean Water Act of 1977, which did not alter the jurisdictional phrase "waters of the United States."[70]

The original version of the Clean Water Act of 1977 introduced in the House would have limited the Corps' jurisdiction, and an amendment proposed in the Senate sought similar limitations, but the original Senate version, which generally retained the existing definition of "navigable waters," was adopted in conference and passed into law. The Clean Water

[68] *See* 33 C.F.R. §323.2(a) (1978).

[69] *See e.g.* S. 867, 95th Cong. (1977) (amending the definition of "navigable waters" to exclude water wholly contained on private property, under the jurisdiction of a state and local government, or which is not susceptible to use as a means to transport commerce); H.R. 3199, 95th Cong. (1977) (redefining "navigable waters" as navigable-in-fact waters and adjacent wetlands).

[70] *See* P.L. 95-217, 91 Stat. 1566 (1977).

Act of 1977, as enacted, contained certain exemptions from Section 404 permitting for "normal farming, silviculture, ... ranching[,]" and other activities.

The original version of the Clean Water Act of 1977 introduced in the House would have limited the Corps' jurisdiction,[71] and an amendment proposed in the Senate sought similar limitations,[72] but the original Senate version, which generally retained the existing definition of "navigable waters," was adopted in conference and passed into law.[73] The Clean Water Act of 1977, as enacted, contained certain exemptions from Section 404 permitting for "normal farming, silviculture, ... ranching[,]" and other activities.[74]

Synthesizing Definitions Following the Clean Water Act of 1977

While the 1977 legislation appeared to temporarily resolve some congressional dispute over the reach of the Clean Water Act, disagreement arose between the Corps and EPA over which agency had final authority to determine which waters were subject to Section 404 permit requirements.[75] EPA independently defined the jurisdictional reach of the Clean Water Act as it related to the programs like NPDES and oil pollution prevention,[76] but it incorporated the Corps' definition into its regulations related to Section

---

[71] *See* H.R. 3199, 95th Cong. §16 (1977) (as introduced) (proposing to redefine "navigable waters" as used in Section 404 to "mean all waters which are presently used, or are susceptible to use in their natural condition or by reasonable improvement as a means to transport interstate or foreign commerce shoreward to their ordinary high water mark, including all waters which are subject to the ebb and flow of the tide").

[72] *See* 123 CONG. REC. 26,710-11 (1977) (proposed amendment by Sen. Bentsen).

[73] *See* 123 CONG. REC. 39,187 (1977) (statement of Sen. Muskie) ("The conference bill follows the Senate bill by maintaining the full scope of Federal regulatory authority over all discharges of dredged or fill material into any of the Nation's waters.").

[74] *See* P.L. 95-217, §67, 91 Stat. at 1600 (codified in 33 U.S.C. §1344(f)).

[75] See Michael C. Blumm & Elisabeth Mering, Vetoing Wetland Permits Under Section 404(c) of the Clean Water Act: A History of Inter-Federal Agency Controversy and Reform, 33 U.C.L.A. J. ENVTL. & POLICY 215, 233 (2015).

[76] See supra note 45 and accompanying text.

404 permits.[77] At the same time, however, EPA separately expanded on that definition in an appendix to its Section 404 regulations.[78]

The U.S. Attorney General ultimately intervened in 1979 and provided a legal opinion that EPA has final administrative authority to determine the reach of the term "navigable waters" for purposes of Section 404.[79] The Corps and EPA eventually executed a Memorandum of Agreement in 1989 resolving that EPA would act as the lead agency responsible for developing programmatic guidance and interpretation of the scope of jurisdictional waters, and the Corps would be responsible for most case-specific determinations on whether certain property was subject to Section 404.[80]

Although it took the agencies 10 years after the Attorney General's opinion to formally agree on a division of responsibilities,[81] the Corps and EPA streamlined and harmonized the regulatory definition of "waters of the United States" well before that. In May 1980, EPA issued regulations redefining the term among its consolidated permit requirements,[82] and the Corps adopted EPA's definition in its regulations two years later.[83] The two agencies continued to use this definition (with modifications) until the Clean Water Rule was published in 2015.[84]

---

[77] See Navigable Waters, Discharge of Dredged or Fill Material, 40 Fed. Reg. 41,292, 41,293 (September 5, 1975) (codified at 40 C.F.R. §230.2(b) (1976)).

[78] See id. at 41,297 app. A.

[79] *See* Benjamin R. Civiletti, *Administrative Authority to Construe § 404 of the Federal Water Pollution Control Act*, 43 Op. Att'y Gen. 197, 197-202 (1979), *available at* https://www.epa.gov/sites/production/files/2015-08/documents/civiletti_memo.pdf.

[80] *See* Dep't of the Army & Envtl. Prot. Agency, Mem. of Agreement: Exemptions Under Section 404(F) of the Clean Water Act (1989), available at http://www.usace.army.mil/Portals/2/docs/civilworks/mous/enfmoa.pdf.

[81] For background on the interagency dispute over the administration of Section 404, see Blum m & Mering, *supra* note 75.

[82] Final Rule, Consolidated Permit Regulations, 45 Fed. Reg. 33,290, 33,424 (May 19, 1980) (codified in 40 C.F.R. §122.3 (1981)).

[83] *See* Interim Final Rule for Regulatory Programs of the Corps of Engineers, 47 Fed. Reg. 31,794, 31,810 (July 22, 1982) (codified in 33 C.F.R. §323.2 (1983)). In its initial proposed regulations implementing the Clean Water Act of 1977, the Corps proposed a shorter, three-category definition which excluded "man-made, non-tidal drainage and irrigation ditches" from "waters of the United States." *See* Proposed Rule, Proposal to Amend Regulations for Controlling Certain Activities in Waters of the United States, 45 Fed. Reg. 62,732, 62,747 (September 19, 1980). That proposal was never adopted.

[84] *Compare* Final Rule, Consolidated Permit Regulations, 45 Fed. Reg. at 33,424 (EPA's definition in the Consolidated Regulations issued in 1980) *and* Interim Final Rule for

The Unified Definition of "Waters of the United States"

By 1982, both the Corps and EPA used the following definition of "waters of the United States" in their regulations:

(a) All waters which are currently used, were used in the past, or may be susceptible to use in interstate or foreign commerce, including all waters which are subject to the ebb and flow of tide;

(b) All interstate waters, including interstate "wetlands";[85]

(c) All other waters such as intrastate lakes, rivers, streams (including intermittent streams), mudflats, sandflats, "wetlands," playa lakes, or natural ponds the use, degradation, or destruction of which would affect or could affect interstate or foreign commerce including any such waters:

1. Which are or could be used by interstate or foreign travelers for recreational or other purposes;
2. From which fish or shellfish are or could be taken and sold in interstate or foreign commerce; or
3. Which are used or could be used for industrial purposes by industries in interstate commerce;

(d) All impoundments of waters otherwise defined as waters of the United States under this definition;

(e) Tributaries of waters identified in paragraphs (1)-(4) of this definition;

(f) The territorial seas; and

(g) "Wetlands" adjacent to waters (other than waters that are themselves wetlands) identified in paragraphs (a)-(f) of this definition.[86]

Regulatory Programs of the Corps of Engineers, 47 Fed. Reg. at 31,810 (the Corps' adoption of EPA's definition) *with* Technical Support for the Clean Water Rule, *supra* note 5, at 18 n.1 (EPA's standard definition immediately prior to the issuance of the Clean Water Rule) *and* 33 C.F.R. §328.3(a)(3) (2014) (the Corps' definition prior to the Clean Water Rule).

[85] "Wetlands" were defined as "those areas that are inundated or saturated by surface or ground water at a frequency and duration sufficient to support, and that do support a prevalence of vegetation.... Wetlands generally include swamps, marshes, bogs, and similar areas." *See* Final Rule, Consolidated Permit Regulations, 45 Fed. Reg. at 33,424.

[86] See 40 C.F.R. §122.3 (1981) (EPA's definition); Interim Final Rule for Regulatory Programs of the Corps of Engineers, 47 Fed. Reg. at 31,810 (codified in 33 C.F.R. §323.2 (1983)) (the Corps' definition).

CHANGES IN JURISDICTIONAL WATERS IN THE 1980S

Riverside Bayview Homes

A legal challenge to the Corps' application of "waters of the United States" was reviewed by the Supreme Court for the first time in 1985 in *United States v. Riverside Bayview Homes, Inc.*[87] There, the Corps sought to enjoin a property owner from discharging fill material on his wetlands located 1 mile from the shore of Lake St. Clair in Michigan,[88] a 468-square-mile, navigable-infact lake that forms part of the boundary between Michigan and Ontario, Canada.[89] The Corps argued that, by defining "waters of the United States" to include wetlands that are "adjacent to" other jurisdictional waters, including navigable-in-fact waters like Lake St. Clair, its regulations required the landowner to obtain a Section 404 permit before discharging fill material.[90]

Before the case reached the Supreme Court, the Sixth Circuit concluded that it must construe the Corps' regulatory definition narrowly in order to avoid a potential violation of the Fifth Amendment prohibition on the taking of private property for public use without just compensation.[91] Applying this method of interpretation, the Sixth Circuit construed the Corps' regulations so as not to include the wetlands at issue, and it avoided reaching a decision on whether the Corps' regulations were constitutional.[92]

The Supreme Court reversed.[93] Although it acknowledged that on a "purely linguistic level" it may seem unreasonable to classify *lands*, wet or otherwise, as *waters*, the Supreme Court called such a plain language

---

87 See 474 U.S. 121 (1985).

88 See id. at 124-25; see also United States v. Riverside Bayview Homes, 729 F.2d 391, 392 (6th Cir. 1984) (describing the wetland property at issue).

89 See Lake Saint Clair, *Encyclopedia Britannica*, https://www.britannica.com/place/Lake-Saint-Clair-lake-NorthAmerica.

90 See Riverside Bayview Homes, Inc., 474 U.S. at 124-25.

91 *See Riverside Bayview Homes*, 729 F.2d at 397-98; *see also* U.S. CONST. amend. V ("nor shall private property be taken for public use, without just compensation").

92 *See Riverside Bayview Homes*, 729 F.2d at 397-98.

93 *Riverside Bayview Homes, Inc.*, 474 U.S. at 139.

approach "simplistic."[94] Further, it rejected the lower courts' concerns over the constitutionality of the Corps' regulations as "spurious."[95] Instead of applying a narrow approach to avoid constitutional implications, the Court gave deference to the Corps' position, and concluded that because "[w]ater moves in hydrological cycles" rather than along "artificial lines," it was reasonable for the Corps to conclude that "adjacent wetlands are inseparably bound up with the 'waters' of the United States...."[96]

The Court also cited legislative history from the passage of the Clean Water Act and the amendments in 1977—in which the term "adjacent wetlands" was added to the statute[97]—as support for its conclusion that Congress intended for the Clean Water Act to have a broad jurisdictional reach which included the adjacent wetlands at issue.[98] In concluding that adjacent wetlands could reasonably be covered, however, the Court also emphasized that it was "not called upon to address the question of the authority of the Corps to regulate discharges of fill material into wetlands that are not adjacent to bodies of open water, ... and we do not express any opinion on that question."[99]

The Migratory Bird Rule and Other Adjustments to "Waters of the United States

Following *Riverside Bayview Homes*, the Corps and EPA engaged in rulemaking in which they interpreted the Clean Water Act to govern all

---

[94] *See id.* at 132. This comment appears to be in response to the Sixth Circuit's statement that "[t]he language of the [Clean Water Act] makes no reference to 'lands' or wetlands' or flooded areas at all." *Riverside Bayview Homes*, 729 F.2d at 397.

[95] *See Riverside Bayview Homes, Inc.*, 474 U.S. at 129.

[96] *Id.* at 133-34.

[97] *See* Clean Water Act of 1977, *See* P.L. 95-217, 91 Stat. 1566, 1601 (1977) (codified in 33 U.S.C. §1344(g)(1).

[98] See *Riverside Bayview Homes, Inc.*, 474 U.S. at 132-34. In a later decision, *Solid Waste Agency of Northern Cook Cnty. v. U.S. Army Corps of Engineers*, the Supreme Court stated that its decision in *Riverside Bayview Homes* "was based in large measure upon Congress' unequivocal acquiescence to, and approval of, the Corps' regulations interpreting the CWA to cover wetlands adjacent to navigable waters." 531 U.S. 159, 180-81 (2001); *see also infra* "*SWANCC*."

[99] *Id.* at 131 n.8.

waters which were used or may have been used by migratory birds crossing state lines.[100] The agencies did not redefine "waters of the United States" through this interpretation, which came to be known as the Migratory Bird Rule, but instead stated the Migratory Bird Rule was a "clarification" of the existing regulatory definition.[101]

The agencies also continued to adjust their interpretation of the definition of "waters of the United States" in the late 1980s by, among other things, excluding nontidal drainage and irrigation ditches, artificial lakes or ponds used for irrigation and stock watering, reflecting pools, and swimming pools.[102] In 1993, the agencies jointly revised their regulations to exclude "prior converted cropland"—areas that were previously drained and converted to agricultural use—from jurisdictional waters.[103]

Competing Wetland Manuals and Congressional Intervention Through Appropriations

In addition to disputes over the textual *definition* of "waters of the United States," disagreement surrounding the technical standards used to *delineate* the boundaries of jurisdictional waters, particularly wetlands, arose in the late 1980s.[104] The Corps issued the first wetlands delineation manual in 1987 (1987 Manual),[105] but EPA published its own manual the

---

[100] *See* Final Rule for Regulatory Programs of the Corps of Engineers, 51 Fed. Reg. 41,206, 41,217 (November 13, 1986); Final Rule: Clean Water Act Section 404 Program Definitions and Permit Exemptions, 53 Fed. Reg. 20,764, 20,765 (June 6, 1988). Although it did not adopt the Migratory Bird Rule in published rulemaking until 1988, EPA. began utilizing it in 1985 before the Court's decision in *Riverside Bayview Homes* was rendered. *See* Mem. from Francis Blake, Gen. Counsel, Envtl. Prot. Agency, to Richard E. Sanderson, Acting Assistant Admin., Office of External Affairs, Envtl. Prot. Agency on Clean Water Act Jurisdiction Over Isolated Waters (September 13, 1985), 1985 WL 195307, at *2.

[101] *See* Final Rule for Regulatory Programs of the Corps of Engineers, 51 Fed. Reg. at 41,217.

[102] *See* Final Rule for Regulatory Programs of the Corps of Engineers, 51 Fed. Reg. at 41,217; Final Rule: Clean Water Act Section 404 Program Definitions and Permit Exemptions, 54 Fed. Reg. at 20,765.

[103] *See* Clean Water Act Regulatory Programs; Final Rule, 58 Fed. Reg. 45,008, 45,031, 45,036-37 (August 25, 1993).

[104] *See Ralph E. Heimlic et al., Wetlands and Agriculture, Private Interests and Public Benefits, AER*-765, 11 (1998).

[105] U.S. Army, Corps of Eng'rs, Technical Report Y-87-1, Wetlands Delineation Manual (1987), *available at* http://www.cpe.rutgers.edu/Wetlands/1987-Army-Corps-Wetlands-Delineation-Manual.pdf. The Corps also created regional supplements to the Wetlands Delineation Manual which are not discussed in this report. *See Regional Supplements to*

following year which utilized an alternative technical analysis.[106] Differences among these and other wetlands manuals led to the preparation of an interagency Federal Manual for Identifying and Delineating Jurisdictional Wetlands in January 1989 (Federal Manual).[107]

Some observers criticized aspects of the Federal Manual, including the methodology it employed for identifying and delineating jurisdictional waters.[108] Some also argued that the Federal Manual improperly expanded the scope of federal regulations of wetlands.[109] Disagreements ultimately led to congressional action in 1991 in the form of appropriations legislation that prohibited the Corps from using funds to identify jurisdictional waters using the Federal Manual.[110] The following year, Congress mandated that the Corps use the 1987 Manual until a new manual was published after public notice and comment.[111] The interagency group proposed revisions to the Federal Manual, which received over 100,000 comments,[112] but that

---

Corps Delineation Manual, USACE.ARMY.MIL (last visited June 27, 2016, 4:59 p.m.), http://www.usace.army.mil/Missions/Civil-Works/Regulatory-Program-and-Permits/reg_supp/.

[106] *See* Heimlic et al., *supra* note 104, at 12 (outlining history of wetlands delineation manuals); *see also James s. Wakeley, Developing a "Regionalized" Version of the Corps of Engineers Wetlands Delineation Manual:* Issues and Recommendations, ERC/CEL TR-02-20, 2 (2001) (describing differences in manuals).

[107] *See U.S. Dep't of the Interior, Fish and Wildlife Serv., et al., Federal Manual for Identifying and Delineating Jurisdictional Wetlands* (1989).

[108] *See* 1989 "Federal Manual for Identifying and Delineating Jurisdictional Wetlands"; Proposed Revisions, 56 Fed. Reg. 40,446 (proposed August 14, 1991) (discussing comments to the Federal Manual); Richard H. McNeer, *Nontidal Wetlands Protection in Maryland and Virginia*, 41 MD. L. REV. 105, 113 (1992) (stating the federal manual "was widely criticized for extending federal jurisdiction to areas that are rarely wet").

[109] *See* Heimlic et al., *supra* note 104, at 12; *see also Wakeley*, *supra* note 106, at 3 ("[T]he 1989 Federal manual generated almost immediate opposition from groups that believed that the manual expanded the Federal government's regulatory authority into lands previously considered to be non-jurisdictional.").

[110] *See* Energy and Water Development Appropriations Act, 1992, P.L. 102-104, 105 Stat. 510, 518.

[111] *See* Energy and Water Development Appropriations Act, 1993, P.L. 102-377, 106 Stat. 1315, 1324-25. Although the appropriations legislation did not reference EPA, EPA agreed to cease using the Federal Manual and use the 1987 Manual in order to create consistency among the agencies' programs. *See* Clean Water Act Regulatory Programs; Final Rule, Memorandum of Agreement Concerning the Determination of the Geographic Scope of the Section 404 Program, 58 Fed. Reg. 4995 (January 19, 1993).

[112] *See* 1989 "Federal Manual for Identifying and Delineating Jurisdictional Wetlands"; Proposed Revisions, 56 Fed. Reg. at 40,446. For background on the opposition to the proposed revisions to the federal manual, see Flint B. Ogle, Comment, *The Ongoing Struggle Between Private Property Rights and Wetlands Regulation: Recent Developments and*

proposal was never finalized, and no interagency wetlands manual was created.[113]

JUDICIALLY IMPOSED LIMITATIONS BEGINNING IN THE LATE 1990S

In contrast to the agencies' attempt to align jurisdictional waters with what they interpreted to be outer reaches of the Commerce Clause in the 1980s, a series of court cases beginning in the late 1990s resulted in the Corps and EPA modifying their interpretation of the phrase "waters of the United States." For much of the 20th century,[114] the Supreme Court broadly construed the Commerce Clause to give Congress discretion to regulate activities which "affect" interstate commerce, so long as its legislation was "reasonably" related to achieving its goals of regulating interstate commerce.[115] In the 1995 case of *United States v. Lopez*, however, the

---

Proposed Solutions, 64 U. COLO. L. REV. 573, 595-96 (1993) and David M. Forman, Comment, *Economic Developments Versus Environmental Protection: Executive Oversight and Judicial Review of Wetland Policy*, 15 HAW. L. REV. 23, 48-49 (1993).

[113] *See* Heimlic et al., *supra* note 104, at 12. EPA and other agencies have continued to publish wetland guidance documents. *See, e.g., Envtl. Prot. Agency, Guiding Principles for Constructed Treatment Wetlands: Providing for Water Quality and Wildlife Habitat* (2000). Because wetland delineation on agricultural properties implicates the Food Security Act and the jurisdiction of Natural Resources Conversation Service within the Department of Agriculture, a separate wetlands delineation manual is used for agricultural lands. *See* WAKELEY, *supra* note 106, at 3.

[114] *See* CRS Report RL32844, *The Power to Regulate Commerce: Limits on Congressional Power*, by Kenneth R. Thomas, at 5 (describing the history of case law interpreting the Commerce Clause including a period of expansive interpretation beginning with *NLRB v. Jones & Laughlin Steel Corporation*, 55 U.S. 1 (1937) and ending with *United States v. Lopez*, 514 U.S. 549 (1995)); *see also* Paul Alexander Fortenberry & Daniel Canton Beck, *Chief Justice Roberts—Constitutional Interpretations of Article III and the Commerce Clause: Will the "Hapless Toad" and "John Q. Public" Have any Protection in the Roberts Court?*, 13 U. *Balt. J. Envtl.* L. 55, 74 ("For roughly sixty years [before *United States v. Lopez*], the Supreme Court had broadly construed Congress's Commerce Clause power.").

[115] *See* United States v. Darby, 312 U.S. 100, 118 (1941) (approving that legislation related to working conditions and stating "[t]he power of Congress over interstate commerce ... extends to those activities intrastate which so affect interstate commerce ... to make regulation of them appropriate means to the attainment of a legitimate end[.]"); Wickard v. Filburn, 317 U.S. 111, 124 (1942) (upholding regulations on price of wheat and stating that even if the regulated "activity be local and though it may not be regarded as commerce, it may still, whatever its nature, be reached by Congress if it exerts a substantial

Supreme Court struck down a federal statute for the first time in more than 50 years based purely on a finding that Congress exceeded its powers under the Commerce Clause.[116]

In *Lopez*, the Court held the Commerce Clause did not provide a constitutional basis for federal legislation criminalizing possession of a firearm in a school zone because the law neither regulated a commercial activity nor contained a requirement that the firearm possession be connected to interstate commerce.[117] The Court revisited its prior Commerce Clause cases and sorted Congress's commerce power into three categories: (1) regulation of channels of commerce, (2) regulation of instrumentalities of commerce, and (3) regulation of economic activities which not only affect but "substantially affect" interstate commerce.[118] *Lopez* set the backdrop for a series of three major opinions limiting federal jurisdiction under the Clean Water Act.

United States v. Wilson

The United States Court of Appeals for the Fourth Circuit issued the first in the series of decisions limiting the jurisdictional reach of the Clean Water Act in 1997.[119] Following a seven-week jury trial in *United States v. Wilson*, three defendants were convicted of violating Section 404[120] for knowingly discharging fill material into wetland property located approximately 10 miles from the Chesapeake Bay and 6 miles from the Potomac River in Maryland.[121] On appeal, the defendants challenged their conviction on the grounds that the portion of Corps' regulatory definition of "waters of the United States"—which included all "waters ... the use,

---

economic effect on interstate commerce"); *see also* CRS Report RL32844, *supra* note 114, at 5 (describing history of Commerce Clause cases).

116 514 U.S. 549 (1995).

117 *See id.* at 551.

118 *See id.* at 558-59.

119 United States v. Wilson, 133 F.3d 251 (4th Cir. 1997). In *Wilson*, the three-judge panel unanimously agreed that the convictions in the district court should be reversed and remanded for a new trial, and a two-judge majority concluded that a portion of the Corps' regulatory definition of "waters of the United States" exceeded the statutory authorization of the Clean Water Act. *See id.* at 257 (Niemeyer & Payne, JJ. joining in part II of the opinion).

120 *See* 33 U.S.C. §§1319(c), 1311(a).

121 *See Wilson*, 113 F. 3d at 254, 256.

degradation or destruction of which *could affect* interstate or foreign commerce"—exceeded the Corps' statutory authority in the Clean Water Act and Congress's constitutional authority in the Commerce Clause.[122]

Relying in part on the holding in *Lopez*, the Fourth Circuit agreed with a portion of the defendants' arguments and ordered a new trial.[123] The court reasoned that, under *Lopez*, the regulated conduct must "substantially affect" interstate commerce in order to invoke the Commerce Clause power, and therefore the Corps exceeded its authority by regulating waters which "could affect" interstate commerce without regard to whether there was any actual effect, substantial or otherwise.[124] And although the Fourth Circuit strongly suggested that the Corps' assertion of jurisdiction exceeded the constitutional grant of authority under the Commerce Clause,[125] it ultimately invalidated the challenged portion of the regulations solely on the ground that it exceeded the congressional authorization under the Clean Water Act.[126]

As *Wilson* never reached the Supreme Court,[127] it was only binding precedent in the Fourth Circuit,[128] and the stricken language remained in the regulations of the Corps and EPA until the release of the Clean Water Rule.[129]

The Corps' 2000 Guidance in Response to Wilson

Although the Corps did not modify its regulatory definition of "waters of the United States" in response to *Wilson*, it did publish guidance in

---

[122] *Id.* at 256-57. (quoting 33 C.F.R. §328.3(a)(3)(1993)) (emphasis in opinion but not in regulation).

[123] *See id.* at 255-57.

[124] *See id.*

[125] *See id.* at 257 ("Were this regulation a statute duly enacted by Congress, it would present serious constitutional difficulties, because, at least at first blush, it would appear to exceed congressional authority under the Commerce Clause.").

[126] *See id.*

[127] Following remand, one defendant pled guilty to a single felony county, which it later unsuccessfully attempted to vacate. *See* United States v. Interstate Gen. Co., 152 F. Supp. 2d 843 (D. Md. 2001), *aff'd sub nom.*, United States v. Interstate Gen. Co., L.P., 39 F. App'x 870 (4th Cir. 2002).

[128] *See* Bonner v. City of Prichard, 661 F.2d 1206, 1209 (11th Cir. 1981) ("the decisions of one circuit are not binding on other circuits"); Duran-Quezada v. Clark Constr. Grp., LLC, 582 Fed. Appx. 238, 239 (4th Cir. 2014) ("the decisions of other circuits are not binding").

[129] *See* 33 C.F.R. §328(a)(3) (2014); 40 C.F.R. §122.2 (2014).

March 2000 on the effect of the decision on its Section 404 jurisdiction.[130] The Corps explained that, *within the Fourth Circuit only*, "isolated waters" must be shown to have an *actual* connection to interstate or foreign commerce.[131] "Isolated waters," in Clean Water Act parlance, are waters that are not navigable-in-fact, not interstate, not tributaries of the foregoing, and not hydrologically connected to such waters—but whose use degradation or destruction could affect interstate commerce.[132]

The 2000 guidance also provided clarification on certain nontraditional waters that the Corps considered part of the "waters of the United States." Jurisdictional waters, the Corps explained, included both *intermittent streams*, which have flowing water supplied by groundwater during certain times of the year, and *ephemeral streams*, which have flowing water only during and for a short period after precipitation events.[133] Drainage ditches constructed in jurisdictional waters were also deemed to be subject to the Clean Water Act except when the drainage was so complete that it converted the entire area to dry land.[134]

SWANCC

In 2001, the Supreme Court took up another challenge to the jurisdictional reach of the Clean Water Act in *Solid Waste Agency of Northern Cook County v. U.S. Army Corps of Engineers* (*SWANCC*), revisiting the issue for the first time since its 1995 decision in *Riverside Bayview Homes*. In *SWANCC*, the Court evaluated whether Clean Water Act jurisdiction extended to an abandoned sand and gravel pit which contained

---

[130] *See* Final Notice of Issuance and Modification of Nationwide Permits, 65 Fed. Reg. 12,818 (March 9, 2000).

[131] *See id.* at 12,824 (emphasis added).

[132] *See* CRS Report RL33263, *The Wetlands Coverage of the Clean Water Act (CWA): Rapanos and Beyond*, by Claudia Copeland and Alexandra M. Wyatt, at 2-3.

[133] *See* Final Notice of Issuance and Modification of Nationwide Permits, 65 Fed. Reg. at 12,823, 12,897-98. Under the 2000 Guidance, ephemeral streams must have an ordinary high water mark to be jurisdictional. *Id.* at 12,823.

[134] *See id.* at 12,823.

water that had become a habitat for migratory birds.[135] Citing the legislative history of the 1972 amendments and the Clean Water Act of 1977, the Corps had argued that the Clean Water Act can extend to such isolated waters under the Migratory Bird Rule.[136]

In a 5-4 ruling, the Court rejected the Corps' position, and held that the Corps' assertion of jurisdiction over isolated waters based purely on their use by migratory birds exceeded its statutory authority.[137] The *SWANCC* Court's conclusion was informed, in part, by *Lopez* and another landmark Commerce Clause decision issued five years later, *United States v. Morrison*,[138] in which the Court held that Congress lacked constitutional authority under the Commerce Clause to enact portions of the Violence Against Women Act.[139] In light of this jurisprudence, the *SWANCC* Court concluded that allowing the Corps to assert jurisdiction under the Migratory Bird Rule raised "serious constitutional questions" about the limits of Congress's authority and "would result in significant impingement of States' traditional and primary power of land and water use."[140] Rather than interpret the Clean Water Act in a way that would implicate these "significant constitutional and federalism questions[,]" the Court concluded that Congress's use of the phrase "navigable waters" in the Clean Water Act "has at least the import of showing us what Congress had in mind for enacting the [Act]: its traditional jurisdiction over waters that were or had been navigable in fact or which could reasonably be made so."[141] Based on this reading, the Court concluded that Congress did not intend to invoke the outer limits of the Commerce Clause in the Clean Water Act, and the Corps could not rely on the Migratory Bird Rule as a basis for jurisdiction.[142]

---

[135] *See* 531 U.S. 159, 162 (2001).

[136] *See id* at 168-70.

[137] *Id.* at 173-74.

[138] 529 U.S. 598, 627 (2000) (affirming decision holding that Congress lacked constitutional authority to enact 42 U.S.C. §13981 (2000)).

[139] *See SWANCC*, 531 U.S. at 173 (citing *Lopez* and *Morrison* and stating "[t]wice in the past six years we have reaffirmed the proposition that the grant of authority to Congress under the Commerce Clause, though broad, is not unlimited.").

[140] *See id.* at 173-74.

[141] *See id.* at 172-74.

[142] *See id.*

In contrast to *Riverside Bayview Homes*, the *SWANCC* Court focused less on the legislative history of the Clean Water Act, and instead emphasized the Corps' original interpretation of the 1972 amendments in which it limited its jurisdiction to navigable-in-fact waters.[143] Although the *Riverside Bayview Homes* Court found that classical "navigability" was of "limited import" in determining Clean Water Act jurisdiction,[144] the *SWANCC* Court distinguished that case as focused on "wetlands adjacent to navigable waters."[145] The ponds which formed in the abandoned gravel pits in *SWANCC* were "*not* adjacent to open water[,]" and therefore lacked the requisite "significant nexus" to traditionally navigable waters necessary for jurisdiction under the Clean Water Act.[146]

SWANCC did not go as far as the Fourth Circuit, however, in striking down an entire subsection of the definition of "waters of the United States." It limited its holding to the Migratory Bird Rule, which was described in the Corps' guidance as an effort to "clarify" the definition.[147] But while its direct holding was arguably narrow, *SWANCC*'s rationale was much broader and called into question whether the Corps and EPA could assert jurisdiction under the Clean Water Act over many wholly intrastate isolated waters.[148] The relationship between *SWANCC*'s limited holding and the Court's broader rationale generated considerable litigation over the scope of the Clean Water Act.[149]

---

[143] *See id.* at 168 (quoting 33 C.F.R. §209.120(d)(1) (1974)).
[144] United States v. Riverside Bayview Homes, Inc. 474 U.S. 121, 133 (1985).
[145] *See SWANCC*, 531 U.S. at 167-68.
[146] *See id.* at 167 (emphasis in original).
[147] *See SWANCC*, 531 U.S. at 174.
[148] *See* Advanced Notice of Proposed Rulemaking on the Clean Water Regulatory Definitions of "Waters of the United States," 68 Fed. Reg. 1991, 1996 (January 15, 2003) (discussing "uncertainties after *SWANCC* concerning jurisdiction over isolated waters that are both intrastate and non-navigable"). Justice Stevens wrote in his dissent that *SWANCC* precluded jurisdiction "over all waters except for actually navigable waters, their tributaries, and wetlands adjacent to each." *See SWANCC*, 531 U.S. at 176-77 (Stevens, J., dissenting).
[149] *See* Robert R. Rm. Verchick, *Toward Normative Rules for Agency Interpretation: Defining Jurisdiction Under the Clean Water Act*, 55 ALA. L. REV. 845, 846 (2004) (describing *SWANCC* as sending the Corps into a "tailspin" of litigation).

Agency Guidance in Response to SWANCC

The general counsels for the Corps and EPA added their voices to the post-*SWANCC* debate in a joint memorandum issued on the last full day of the Clinton Administration, January 19, 2001.[150] Combining the "significant nexus" language from *SWANCC* with the existing regulatory definition of "waters of the United States," the agencies concluded that they could continue to exercise jurisdiction over isolated waters so long as the use, degradation, or destruction of those waters could affect other "waters of the United States."[151] The potential effect on or degradation on existing jurisdictional waters, the agencies reasoned, established the "significant nexus" mentioned in *SWANCC.*[152]

In January 2003, the Corps and EPA issued a notice of proposed rulemaking regarding how field staff should address jurisdictional issues in the Clean Water Act and which contained a revised joint memorandum on the effect of *SWANCC.*[153] That proposed rulemaking effort was later abandoned, leaving unanswered questions over the agencies' jurisdiction over isolated waters after *SWANCC.*[154] These uncertainties caused the Corps and EPA to shift their attention to alternative bases for jurisdiction in defining "waters of the United States"—such as "adjacent wetlands"—and set the stage for the Supreme Court's next encounter with a Clean Water Act jurisdictional dispute in *Rapanos v. United States*.

---

150 Joint Memorandum from Gary S. Guzy, General Counsel, U.S. Envtl. Prot. Agency, and Robert M. Andersen, Chief Counsel, U.S. Army Corps of Eng'rs on Supreme Court Rule Concerning CWA Jurisdiction Over Isolated Waters (January 19, 2001) [hereinafter 2001 Joint Memorandum], *available* at https://www.environment.fhwa.dot.gov/ecosystems/laws_swepacoe.asp.

151 *See id.* at 3.

152 *See id.* ("With respect to waters that are isolated, intrastate, and nonnavigable—jurisdiction may be possible if their use, degradation, or destruction could affect other 'waters of the United States,' thus establishing a significant nexus between the water in question and other 'waters of the United States[.]").

153 *See* Advanced Notice of Proposed Rule Making on the Clean Water Regulatory Definitions of "Waters of the United States," 68 Fed. Reg. at 1991, 1995 app. A.

154 *See* Rapanos v. United States, 547 U.S. 715, 758 (2006) (Kennedy, J., concurring) ("The proposed rulemaking went nowhere.").

Rapanos

> **The Competing Approaches Following *Rapanos***
>
> *The Plurality's Bright-Line Rule*: Writing for a four-Justice plurality, Justice Scalia adopted the bright-line rule that the word "waters" in "waters of the United States" means only "relatively permanent, standing or continuously flowing bodies of water"—that is, streams, rivers, and lakes.[155] Wetlands could also be included, but only when they have a "continuous surface connection" to other "waters of the United States."[156]
>
> *Justice Kennedy's "Significant Nexus" Test:* In a separate concurring opinion, Justice Kennedy concluded that the Clean Water Act requires a more malleable approach: the Corps should determine, on a case-by-case basis, whether the water in question possesses a "significant nexus" to waters that are navigable-in-fact.[157] For wetlands, a significant nexus exists when the wetland, either alone or in connection with similarly situated properties, significantly impacts the chemical, physical, and biological integrity of a traditionally navigable waterbody.[158]

Rapanos involved a consolidation of two cases on appeal from the Sixth Circuit—*Rapanos*[159] and *Carabell*[160]—both of which involved disputes over the breadth of the Clean Water Act's jurisdiction over "adjacent wetlands." In *Carabell*, landowners challenged whether Section 404 jurisdiction extends to "wetlands that are hydrologically isolated from any of the 'waters of the United States[,]'"[161] and *Rapanos* presented the similar question of whether this jurisdiction includes nonnavigable wetlands "that do not even abut a navigable water."[162] In both cases, which are collectively referred to as *Rapanos*, the Sixth Circuit upheld the Corps' assertion of jurisdiction over the wetland property at issue.[163]

---

155 *See Rapanos*, 547 U.S. at 739.

156 *See id.* at 742.

157 *See id.* at 782 (Kennedy, J., concurring).

158 *See id.* at 780.

159 Rapanos v. United States, 376 F.3d 629 (6th Cir. 2004), *cert. granted*, 546 U.S. 932-33 (2005).

160 Carabell v. U.S. Army Corps of Eng'rs, 391 F.3d 704 (6th Cir. 2004), *cert. granted*, 546 U.S. 932-33 (2005).

161 Questions Presented, U.S. Army Corps of Eng'rs, No. 04-1384, 547 U.S. 715 (2006), available at http://www.supremecourt.gov/qp/04-01384qp.pdf.

162 Questions Presented, Rapanos v. United States, No. 04-1034, 547 U.S. 715 (2006), available at http://www.supremecourt.gov/qp/04-01034qp.pdf.

163 *See Rapanos*, 547 U.S. at 729-30.

Many anticipated that *Rapanos* would provide clarity on the disputes following *SWANCC*.[164] And although a majority of five Justices agreed that the Sixth Circuit decision was flawed, they were not able to agree on a single, underlying standard which would govern future jurisdictional disputes. Instead, a four-Justice plurality opinion, authored by Justice Scalia, and an opinion by Justice Kennedy, writing only for himself, proposed two alternative tests for determining jurisdictional waters.

Lower Courts' Response to Rapanos

With no controlling rationale from the majority, lower courts interpreting Rapanos struggled with the question of what analysis to apply in Clean Water Act jurisdictional disputes.[165] Of the nine circuits which have addressed the issue thus far,[166] all have applied Justice Kennedy's significant nexus test either alone or in combination with the plurality's test, and none have applied the plurality approach alone.[167] Still, some

---

[164] *See, e.g.*, Gregory T. Broderick, *From Migratory Birds to Migratory Molecules: The Continuing Battle over the Scope of Federal Jurisdiction Under the Clean Water Act*, 30 colum. J. Envtl. L. 473, 522 (2004) ("With the lower courts in conflict and the political branches unable to move on this important question [of CWA jurisdiction,] only the Supreme Court can fix the problem."); CRS Report RL33263, *supra* note 132, at 5 ("For many who had waited so long to have 'waters of the United States' clarified, the *Rapanos* decision ... was a disappointment.").

[165] In his brief concurrence, Chief Justice Roberts predicted difficulties in implementing *Rapanos*, stating the following: "It is unfortunate that no opinion commands a majority of the Court on precisely how to read Congress' limits on the reach of the Clean Water Act. Lower courts and regulated entities will not have to feel their way on a case-by-case basis." *See id.* 547 U.S. at 758 (Roberts, C.J., concurring). For a more detailed analysis of lower courts' varying interpretations of *Rapanos*, see CRS Report RL33263, *supra* note 132, at 7-8 and Technical Support for the Clean Water Rule, *supra* note 5, at 40-47.

[166] *See* Precon Dev. Corp. v. U.S. Army Corps of Eng'rs, 633 F.3d 278 (4th Cir. 2011); United States v. Donovan, 661 F.3d 174 (3d Cir. 2011), *cert. denied*, 132 S. Ct. 2409 (2012); United States v. Bailey, 571 F.3d 791 (8th Cir. 2009); United States v. Cundiff, 555 F.3d 200 (6th Cir.), *cert. denied*, 130 S. Ct. 74 (2009); United States v. Lucas, 516 F.3d 316 (5th Cir.), *cert. denied*, 555 U.S. 822 (2008); United States v. Robison, 505 F.3d 1208 (11th Cir. 2007), *cert. denied sub nom* McWane v. United States, 555 U.S. 1045 (2008); N. Cal. River Watch v. City of Healdsburg,496 F.3d 993 (9th Cir. 2007) (superseding the original opinion published at 457 F.3d 1023 (9th Cir. 2006)), *cert. denied*, 552 U.S. 1180 (2008); United States v. Johnson, 467 F.3d 56 (1st Cir. 2006), *cert. denied*, 552 U.S. 948 (2007); United States v. Gerke Excavating, Inc., 464 F.3d 723 (7th Cir. 2006), *cert. denied*, 552 U.S. 810 (2007).

[167] When a majority of the Supreme Court agrees only on the outcome of a case and not on the ground for that outcome, the holding of the Court which lower courts must follow "may be viewed as that position taken by those Members who concurred in the judgments on the

courts and observers have criticized the significant nexus test as vague and difficult to implement.[168]

Agency Guidance in Response to Rapanos

The Corps and EPA offered their own interpretation of *Rapanos* through guidance to field officers in 2007,[169] which was revised and replaced after public comment in 2008.[170] The 2008 guidance adopts the view taken by some lower courts[171] that jurisdiction exists over any water body that satisfies *either* the plurality approach *or* the significant nexus test.[172] The agencies further deconstructed the jurisdictional analysis into

narrowest grounds...." Marks v. United States, 430 U.S. 188, 193 (1977) (quoting Gregg v. Georgia, 428 U.S. 153, 169 n.15 (1976) (opinion of Stewart, Powell, and Stevens, JJ.)). Some courts have concluded that Justice Kennedy's "significant nexus" test is the narrowest ruling to be derived from *Rapanos*. *See Gerke Excavating, Inc.* 464 F.3d at 725 ("[A]s a practical matter the Kennedy concurrence is the least common denominator[.]"); *Robison*, 505 F.3d at 1222 ("[P]ursuant to *Marks*, we adopt Justice Kennedy's 'significant nexus' test as the governing definition of 'navigable waters' under *Rapanos*."). Others courts interpreted *Marks* to conclude "that two new tests should apply." *Donovan*, 661 F.3d at 183-84; *accord Bailey*, 571 F.3d at 799; *Johnson,* 467 F.3d at 66.

For more background on the differing circuit court approaches, see CRS Report RL33263, *supra* note 132, at 7-8 and Technical Support for the Clean Water Rule, *supra* note 5, at 40-47.

168 *See, e.g.*, United States v. Chevron Pipe Line Co., 437 F. Supp. 2d 605 (N.D. Tex. 2006) ("This test leaves no guidance on how to implement its vague, subject centerpiece."); Annie Snider, *The Two Words that Rewrote American Water Policy*, POLITICO (May 25, 2016), http://www.politico.com/agenda/story/2016/05/obama-wotus-wetlands-rulesupreme-court-000131 ("[A]s definitive as those words [significant nexus] sound, the real problem was—and still is— that nobody has ever known quite what they were supposed to mean."); Lowell M. Rothschild, *The Practical Application of the Significant Nexus Test: The Final Waters of the US Rule*, NAT. L. REV. (June 8, 2015), https://www.natlawreview.com/article/practical-application-significant-nexus-test-final-waters-us-rule ("[T]he significant nexus test ... is an ambiguous, case-by case test.").

169 Mem. from Envtl. Prot. Agency & Dep't of the Army on Clean Water Act Jurisdiction Following the U.S. Supreme Court's Decision in *Rapanos v. United States & Carabell v. United States* (June 5, 2007), *available at* https://www.epa.gov/sites/production/files/2016-04/documents/rapanosguidance6507.pdf.

170 Revised Mem. from Envtl. Prot. Agency & Dep't of the Army on Clean Water Act Jurisdiction Following the U.S. Supreme Court's Decision in *Rapanos v. United States & Carabell v. United States* (December 2, 2008) [hereinafter 2008 Memorandum], available at https://www.epa.gov/sites/production/files/2016-02/documents/cwa_jurisdiction_followingjapanos120208.pdf.

171 *See* United States v. Donovan, 661 F.3d 174 (3d Cir. 2011), *cert. denied*, 132 S. Ct. 2409 (2012); United States v. Bailey, 571 F.3d 791 (8th Cir. 2009); United States v. Johnson, 467 F.3d 56 (1st Cir. 2006), *cert. denied*, 552 U.S. 948 (2007).

172 2008 Memorandum, *supra* note 170, at 3.

three categories: (1) waters that are categorically jurisdictional; (2) waters that may be deemed jurisdictional on a case-by-case basis; and (3) waters that are excluded from jurisdiction under the Clean Water Act.[173]

Joint Guidance in Response to *Rapanos*

The Corps and EPA issued joint guidance in 2008 in which they reorganized the jurisdictional analysis into three types of waters:

1. *Waters that are categorically "waters of the United States,"* including navigable-in-fact waters, "relatively permanent" tributaries, and wetlands that have a continuous surface connection or unbroken hydrological connection to jurisdictional waters;
2. *Waters that may be deemed "waters of the United States" on a case-by-case basis* upon a finding of a significant nexus with other jurisdictional waters, such as intermittent and ephemeral streams and wetlands that do not meet the criteria above; and
3. *Waterbodies that are excluded from "waters of the United States,"* including swales or gullies and ditches wholly in and draining only upland that do not carry a relatively permanent flow of water.[174]

In 2011, the Corps and EPA sought comments on proposed changes to the 2008 guidance, which the agencies acknowledged would increase the number of waters regulated under the Clean Water Act in comparison to its earlier post-*Rapanos* guidance.[175] According to the agencies, the 2011 draft guidance was focused on protecting smaller waters that feed into larger ones in an effort to keep downstream water safe from upstream pollutants.[176] The potential enlargement of jurisdiction spawned congressional attention, including a letter signed by 41 Senators requesting

---

[173] *See id.* at 4-11.

[174] *See id.*

[175] *See* EPA and Army Corps of Eng'rs. Guidance Regarding Identification of Waters Protected by the Clean Water Act, 76 Fed. Reg. 24,479 (May 2, 2011) ("The agencies believe that under this proposed guidance the number of waters identified as protected by the Clean Water Act will increase compared to current practice....").

[176] *See* CRS Report RL33263, *supra* note 132, at 12 (discussing the 2011 draft guidance).

that the agencies abandon the effort.[177] Prohibitions on funding related to the draft guidance were included in several appropriations bills, but those provisions were never enacted.[178] Instead, the agencies abandoned pursuit of the 2011 draft guidance in favor of their most recent effort at defining the scope of "waters of the United States," the Clean Water Rule.

THE CLEAN WATER RULE

The Clean Water Rule marks the culmination of the Corps' and EPA's effort to clarify the bounds of jurisdictional waters in the wake of *SWANCC* and *Rapanos*.[179] In developing the Rule, the agencies relied on a synthesis of more than 1,200 published and peer-reviewed scientific reports related to the current scientific understanding of the connection or isolation of streams and wetlands relative to large water bodies like rivers, lakes, estuaries, and oceans.[180] After receiving over 1 million comments to a proposed version of the rule, the agencies issued the final Clean Water Rule on May 27, 2015.[181]

---

[177] *See* Letter from Sen. James Inhofe et al. to Lisa P. Jackson & Jo-Ellen Darcy (June 30, 2011), *available at* http://www.epw.senate.gov/public/_cache/files/ec609d07-a036-49e8-a8a0-c46652b479bd/110630-jackson-darcy-cwaguidance.pdf.

[178] See H.R. 4923, 113th Cong. §106; H.R. 2584, 112th Cong. §435; H.R. 6061, 112th Cong. §434; see also CRS Report R43455, *EPA and the Army Corps' Rule to Define "Waters of the United States"*, by Claudia Copeland, at 1-2 (discussing legislative proposals to bar EPA and the Corps from implementing the 2011 proposed guidance or developing regulations based on it).

[179] *See* Proposed Rule, Definition of "Waters of the United States" Under the Clean Water Act, 79 Fed. Reg. 22,188 (proposed April 21, 2014) (discussing background for proposed Clean Water Rule).

[180] *See* Final Rule, Clean Water Rule, Definition of "Waters of the United States," 80 Fed. Reg. 37,054, 37,057 (June 29, 2015) [hereinafter Clean Water Rule].

[181] *See id.* EPA extended the public comment period on two occasions. *See* Proposed Rule; Extension of Comment Period, 79 Fed. Reg. 35,712 (June 24, 2014) (extending public comment period to October 20, 2014); Proposed Rule; Extension of Comment Period, 79 Fed. Reg. 61,590 (October 14, 2014) (extending public comment period to November 14, 2014).

Key Provisions of the Clean Water Rule[182]

- Navigable waters, interstate waters, the territorial seas, or impoundments of such waters are categorically jurisdictional.
- Tributaries—as newly defined in the Clean Water Rule[183]—of traditional navigable waters, interstate waters, and the territorial seas are categorically jurisdictional.
- Waters, including wetlands, lakes, ponds, and "similar waters," that are adjacent to traditional navigable waters, interstate waters, and the territorial seas are categorically jurisdictional.
- Some waters would remain subject to a case-specific evaluation as to whether they have a significant nexus to traditional navigable waters, interstate waters, and the territorial seas.
- A number of waters would be categorically excluded from Clean Water Act jurisdiction, including prior converted cropland, groundwater and certain ditches, and stormwater management systems.

The final version contains the same three-tier structure from the agencies' 2008 joint guidance, identifying waters that are (1) categorically jurisdictional, (2) may be deemed jurisdictional on a case-by-case basis if they have a significant nexus with other jurisdictional waters, and (3) categorically excluded from the Clean Water Act's jurisdiction.[184] A significant impetus behind the Clean Water Rule was an effort to remove the uncertainty for the landowner and administrative burden to the Corps and EPA created when individual waters and wetlands are evaluated on a case-by-case basis.[185] To that end, the new rule increases categorical

[182] *See* Clean Water Rule 80 Fed. Reg. at 37,057-59 (discussing "Major Rule Provisions" and summarizing other elements of the Clean Water Rule); *id.* at 37,104-06 (redefining "waters of the United States" in 33 C.F.R. §329.3) *see also* CRS In Focus IF10125, *Overview of EPA and the Army Corps' Rule to Define "Waters of the United States"*, by Claudia Copeland, at 1. The details of the Clean Water Rule are also discussed in CRS Report R43455, *EPA and the Army Corps' Rule to Define "Waters of the United States"*, by Claudia Copeland.

[183] *See Clean Water Rule*, 80 Fed. Reg. at 37,105.

[184] *See* Clean Water Rule, 80 Fed. Reg. at 37,057.

[185] *See id.* ("This rule replaces existing procedures that often depend on individual, time-consuming, and inconsistent analyses of the relationship between [waters.]"); *What the Clean Water Rules Does*, EPA (May 18, 2016), https://www.epa.gov/cleanwaterrule/what-clean-water-rule-does ("The rule significantly limits the use of case-specific analysis by creating clarity and certainty on protected waters and limiting the number of similarly situated water features.").

determinations as to whether certain property contains "waters of the United States."[186]

Response to the Clean Water Rule

Many reported that the Clean Water Rule was met with controversy,[187] and a Government Accountability Office (GAO) report found that EPA violated publicity or propaganda and antilobbying provisions in prior appropriations acts through its promotion of the Clean Water Rule on social media.[188] Congress also took steps to block its implementation. In January 2016, the Senate and House passed a resolution of disapproval seeking to nullify the Clean Water Rule[189] under the Congressional Review Act;[190] but that resolution was vetoed by the President.[191] On January 21, 2016, a procedural vote in the Senate to override the veto failed.[192]

The Clean Water Rule was scheduled to take effect on August 28, 2015;[193] however, numerous lawsuits were filed soon after it was announced.[194] While EPA and the Corps contend that the Clean Water Rule

---

[186] *See* Clean Water Rule, 80 Fed. Reg. at 37,057 ("The agencies have greatly reduced the extent of waters subject to this individual review...."); Richard M. Glick and Diego Atencio, *"Waters of the United States" Not Quite Clear Yet*, WATER REP., July 15, 2016, at 3 ("The new rule increases categorical jurisdictional determinations, and is intended to minimize the need for case-specific analyses.").

[187] *See, e.g., id.* at 14 ("The rule was controversial even before it was proposed in March 2014, and controversies have persisted since the final rule was issued"); Snider, *supra* note 168 ("When the Obama administration released [the Clean Water Rule] in early summer 2015, it detonated across the American heartland like a bomb.").

[188] *See* U.S. Gov't Accountability Office, GAO B-326944, Environmental Protection Agency—Application of Publicity or Propaganda and Anti-Lobbying Provisions 26 (2015), *available at* http://www.gao.gov/assets/680/674163.pdf.

[189] *See* S.J.Res. 22, 114th Cong. (2016).

[190] *See* 5 U.S.C. §§801-808.

[191] *See* U.S. President (Obama), Veto Message from the President (January 19, 2016), available at https://www.whitehouse.gov/the-press-office/2016/01/19/president-obama-vetoes-sj-22.

[192] *See* U.S. Senate, Roll Call Votes on the Motion to Invoke Cloture on the Veto Message to Accompany S.J.Res. 22 (January 21, 2016), http://www.senate.gov/legislative/LIS/roll_call_lists/roll_call_vote_cfm.cfm?congress=114& session=2&vote=00005#top.

[193] *See* Clean Water Rule, 80 Fed. Reg. at 37,054.

[194] *See* CRS Report RL33263, *supra* note 132, at 15; *see also* Amena H. Saiyid, *Litigation Tracker: Water Rule Challenges Dominate Courts*, BNA, July 26, 2016, http://news.

governs only waters that have historically been covered by the Clean Water Act,[195] its opponents argue it constitutes an unlawful expansion of authority beyond that which is allowed by the act or the Constitution and which is over-burdensome to private landowners.[196]

In one lawsuit, the U.S. Court of Appeals for the Sixth Circuit issued an order staying its implementation as of October 9, 2015.[197] The agencies have since stated that they intend to defend the Clean Water Rule until a final decision is reached by the courts, but that they will apply their prior regulations during the pendency of the stay.[198]

The most recent action related to the Clean Water Rule and jurisdictional waters occurred on July 14, 2016, when the House passed its version of the FY2017 Interior and Environment Appropriations Act.[199] That bill, as passed by the House, would prohibit the use of appropriated funds to adopt or enforce any change to jurisdictional waters beyond those that were in effect on October 1, 2012.[200]

---

bna.com/wrrn/WRRNWB/split_display.adp?fedfid=94728872&vname=wrrnotallissues&wsn=499592500&searchid=28123244&doctypeid=1&type=date&mode=doc&split=0&scm=WRRNWB&pg=0 (summarizing nationwide challenges to the Clean Water Rule in various courts).

195 *See* Envtl. Prot. Agency & U.S. Army Corps of Eng'rs., Factsheet, Clean Water Rule, https://www.epa.gov/sites/production/files/2015-05/documents/fact_sheet_summary_final_1.pdf.

196 *See, e.g.*, Snider, *supra* note 163 (describing opposition and legal challenges to the Clean Water Act); Press Release, U.S. Speaker of the House, Speaker Boehner on the Latest EPA Power Grab (May 27, 2015), http://www.speaker.gov/press-release/speaker-boehner-latest-epa-power-grab (calling the Clean Water Rule a "raw and tyrannical power grab").

197 *See* Ohio v. U.S. Army Corps of Eng'rs., 803 F.3d 804 (6th Cir. 2015). More information on the current status of this litigation is available through CRS Legal Sidebar WSLG1503, *UPDATED: Sixth Circuit Will Hear Challenges to EPA's Clean Water Act Jurisdiction ("Waters of the United States") Rule, but Litigation Uncertainties Remain Unresolved*, by Alexandra M. Wyatt.

198 *See* Mem. from Envtl. Prot. Agency & Dep't of the Army on Administration of Clean Water Programs in Light of the Stay of the Clean Water Rule; Improving Transparency and Strengthening Coordination (November 16, 2015), at 1- 2, *available at* https://www.epa.gov/sites/production/files/2015-11/documents/2015-11-16_signed_cwr_post-stay_coordination_memo.pdf.

199 *See* H.R. 5538, 114th Cong. (2016).

200 *See id.* at §427.

APPENDIX. TABLE CONCERNING MAJOR FEDERAL ACTIONS RELATED TO "WATERS OF THE UNITED STATES" IN THE CLEAN WATER ACT

Table A-1. Major Federal Actions Related to "Waters of the United States" in the Clean Water Act

Date	Event	Source
Oct. 18, 1972	Federal Water Pollution Control Act Amendments of 1972 is enacted	P.L. 92-500, 86 Stat. 816
Feb. 6, 1973	EPA interprets the scope of Clean Water Act jurisdiction in an Office of the General Counsel Memorandum	1973 WL 21937
May 4, 1973	The Corps proposes its first regulations defining "navigable waters" under the Clean Water Act	38 Fed. Reg. 12,217
May 22, 1973	EPA issues regulations defining "navigable waters" under the Clean Water Act[201]	38 Fed. Reg. 13,528
April 1, 1974	The Corps issues final regulations defining "navigable waters" under the Clean Water Act	39 Fed. Reg. 12,115
Mar. 27, 1975	The U.S. District Court for the District of Columbia strikes down the Corps' regulatory definition in *National Resources Defense Council v. Callaway*	392 F. Supp. 685
May 6, 1975	The Corps publishes proposed regulations in response to *Callaway*	40 Fed. Reg. 19,766
July 25, 1975	The Corps publishes final interim regulations revising the definition of "navigable waters"	40 Fed. Reg. 31,320
July 19, 1977	The Corps publishes a final rule defining "waters of the United States"	42 Fed. Reg. 37,122
Dec. 27, 1977	The Clean Water Act of 1977 is enacted	P.L. 95-217, 91 Stat. 1566

[201] For purposes of this table, the Clean Water Act refers to the Federal Water Pollution Control Act Amendments of 1972 and subsequent amendments and related legislation.

Date	Event	Source
Aug. 28. 1975	EPA adopts the Corps' definition of "navigable waters" under the Section 404 program	40 Fed. Reg. 41,292
Sept. 5, 1979	Attorney General Ben Civiletti publishes opinion that EPA has ultimate responsibility to determine jurisdictional waters	43 Op. Att'y Gen. 197
May 19, 1980	EPA publishes a Final Rule defining "waters of the United States"	45 Fed. Reg. 33,290
Sept. 19, 1980	The Corps issues a proposed rule with a definition of "waters of the United States" that continues to differ from EPA	45 FR 62,732
July 22, 1982	The Corps issues an interim final rule adopting EPA's definition of "waters of the United States"	47 FR 31,794
Sept. 13, 1985	EPA's General Counsel writes a memorandum on the applicability of the Migratory Bird Rule	1985 WL 195307
Dec. 4, 1985	The Supreme Court decides United States v. Riverside Bayview Homes, Inc.	474 U.S. 121
Nov. 13, 1986	The Corps issues guidance adopting the Migratory Bird Rule	51 Fed. Reg. 41,206
June 6, 1988	EPA publishes guidance adopting the Migratory Bird Rule	53 Fed. Reg. 20,764
Jan. 1987	The Corps publishes the first wetlands delineation manual	http://www.cpe.rutgers.edu/Wetlands/1987-Army-Corps-Wetlands-Delineation-Manual.pdf
April 1988	EPA publishes a wetlands delineation manual	http://www.ers.usda.gov/media/929243/aer765_002.pdf[202]

[202] The 1988 EPA manual is no longer publically disseminated or available on EPA's website; however, several secondary sources discuss the manual. *See, e.g.*, HEIMLIC ET AL., *supra* note 104, at 11-12; WAKELEY, *supra* note 106, at 3.

Table A-1. (Continued)

Date	Event	Source
Jan. 10, 1989	An interagency group publishes a federal wetlands delineation manual	https://www.fws.gov/northeast/ecologicalservices/pdf/wetlands/interagency%20wetland%20delineation%20manual%201989.pdf
Jan. 19, 1989	The Corps and EPA execute a Memorandum of Agreement regarding their respective jurisdiction under Section 404	http://www.usace.army.mil/Portals/2/docs/civilworks/mous/enfmoa.pdf
Aug. 14, 1991	An interagency group proposes revisions to the 1989 federal wetlands delineation manual	56 Fed. Reg. 40,446
Aug. 17, 1991	Appropriations legislation is passed prohibiting the Corps from using the 1989 federal wetlands delineation manual	P.L. 102-104, 105 Stat. 510
Oct. 2, 1992	Appropriations legislation is passed requiring the Corps to use the 1987 wetlands delineation manual	P.L. 102-377, 106 Stat. 1315
Jan. 19, 1993	EPA ceases to use the 1989 federal wetlands delineation manual, and instead uses the Corps' 1987 manual	58 Fed. Reg. 4995
Aug. 25, 1993	The Corps and EPA revise regulations to exclude "prior converted cropland" from "waters of the United States"	58 Fed. Reg.45,008
Dec. 23, 1997	The Fourth Circuit invalidates a portion of the Corps' definition of "waters of the United States" in United States v. Wilson	133 F.3d 251
Mar. 9, 2000	The Corps publishes guidance on the effect of Wilsonand on other nontraditional "waters of the United States" including ephemeral streams, intermittent streams, and drainage ditches	65 Fed. Reg. 12,818

Date	Event	Source
Jan. 9, 2001	The Supreme Court holds that the Corps cannot exercise jurisdiction over waters or wetlands based solely on the Migratory Bird Rule in Solid Waste Agency of Northern Cook County v. U.S. Army Corps of Engineers (SWANCC)	531 U.S. 159
Jan. 19, 2001	The general counsels of the Corps and EPA issue a joint memorandum on the on the effect of SWANCC	https://www.environment.fhwa.dot.gov/ecosystems/laws_swepacoe.asp
Jan. 15, 2003	The Corps issues proposed rulemaking addressing how field staff should address jurisdictional waters issues, but it is never finalized	68 Fed. Reg. 1991
June 19, 2006	The Supreme Court decides Rapanos v. United States and Carabell v. United States Army Corps of Engineers	547 U.S. 715
June 5, 2007	The Corps and EPA issue joint guidance on the impact of Rapanos	https://www.epa.gov/sites/production/files/2016-04/documents/rapanosguidance6507.pdf
Dec. 2, 2008	The Corps and EPA revise the joint guidance on Rapanos	https://www.epa.gov/sites/production/files/2016-02/documents/cwa_jurisdiction_following_rapanos120208.pdf
May 2, 2011	The Corps and EPA seek comments on proposed guidance which would increase the number of waters regulated under the Clean WaterAct; the proposed guidance is never finalized	76 Fed. Reg. 24,479
April 21, 2014	The Corps and EPA issue the proposed Clean Water Rule	79 FR 22,188
June 29, 2015	The Corps and EPA issue the final Clean Water Rule	80FR 37,053
Oct. 9, 2015	The U.S. Court of Appeals for the Sixth Circuit stays the application off the Clean Water Rule	803 F.3d 804

Table A-1. (Continued)

Date	Event	Source
Nov. 16, 2015	The Corps and EPA issue a joint memorandum regarding their approach to resolving jurisdictional waters issues during the stay of the Clean Water Rule	https://www.epa.gov/sites/production/files/2015-11/documents/2015-11-16_signed_cwr_post-stay_coordination_memo.pdf
Jan. 13, 2016	Senate and House pass resolution of disapproval, S.J.Res. 22, seeking to nullify the Clean Water Rule	S.J.Res. 22, 114th Cong.
Jan. 19, 2016	The President vetoes S.J.Res. 22	https://obamawhitehouse.archives.gov/the-press-office/2016/01/19/president-obama-vetoes-sj-22
Jan. 21, 2016	A procedural vote to override the President's veto of S.J.Res. 22 fails in the Senate	http://www.senate.gov/legislative/LIS/roll_call_lists/roll_call_vote_cfm.cfm?congress=114&session=2&vote=00005#top
July 14, 2016	The House passes its version of the FY2017 Interior and Environment Appropriations Bill, which contains a provision that would prohibit the use of appropriated funds to adopt or enforce a change to jurisdictional waters beyond those that were in effect on October 1, 2012	H.R. 5538, §427, 114thCong.

Source: Congressional Research Service; based on sources cited in this report.

In: Water Quality
Editors: Edward Saunders et al.
ISBN: 978-1-53614-809-1

Chapter 4

CLEAN WATER RULE CHALLENGES MUST BEGIN IN FEDERAL DISTRICT COURTS, SUPREME COURT HOLDS*

Stephen P. Mulligan

In *National Association of Manufacturers (NAM) v. Department of Defense*, the Supreme Court held that legal challenges to the 2015 Clean Water Rule issued by the Environmental Protection Agency (EPA) and Army Corps of Engineers (Corps) must be litigated in multiple federal district courts rather than in a consolidated case in the United States Court of Appeals for the Sixth Circuit (Sixth Circuit). Since October 2015, a Sixth Circuit order had stayed implementation of the Clean Water Rule on a nationwide basis pending the court's determination of its jurisdiction over the challenge to the Rule. But the Supreme Court's *NAM* decision, which requires the Sixth Circuit to dismiss its case, will result in the elimination of that order, and potentially could complicate legal challenges

* This is an edited, reformatted and augmented version of Congressional Research Service, Publication No. LSB10069, dated January 31, 2018.

to the Clean Water Rule and the Trump Administration's effort to rescind and replace it.

BACKGROUND ON *NAM* AND THE CLEAN WATER RULE

Also known as the "Waters of the United States" or WOTUS Rule, the Clean Water Rule attempts to define which waters and wetlands are subject to regulatory requirements in the Clean Water Act. The underlying issue addressed in the Clean Water Rule—the breadth of federal regulatory jurisdiction under the Clean Water Act—has been debated and litigated for more than 40 years, and the Clean Water Rule has been no exception to this trend. After the Corps and EPA announced the Clean Water Rule in 2015, more than 100 parties filed lawsuits challenging its scope and legal authority in federal appellate and district courts across the country. Before any court could address the merits of the claims, however, an impasse arose over what court was the proper forum for the litigation.

Whereas many opponents of the Clean Water Rule argued that their cases should be litigated at the federal district court level (where challenges to agency actions ordinarily begin), the United States contended that the cases fell within the scope of § 509 of the Clean Water Act, which lists seven categories of agency actions that are subject to direct appellate court review. In a split decision issued in December 2015 (discussed here), a panel of the Sixth Circuit adopted the United States' position and held that § 509 gave the federal appellate courts original jurisdiction to hear all challenges to the Clean Water Rule. But, in the latest turn in the long-winding litigation over the Clean Water Act's jurisdiction, the Supreme Court disagreed with the Sixth Circuit and vacated its opinion with instructions to dismiss the case for lack of jurisdiction. "Congress has made clear that rules like the [Clean Water] Rule must be reviewed first in federal district court[,]" wrote Justice Sotomayor in an opinion for the unanimous Court.

The Future of the Legal Challenge to the Clean Water Rule After *NAM*

Although the Supreme Court's *NAM* decision does not resolve the merits of the challenges to the Clean Water Rule, it could have important implications for how the legal process unfolds. In 2015, suits seeking to invalidate the Rule were proceeding in 13 federal district courts, but those courts stayed or dismissed their cases when the Sixth Circuit issued its now-vacated decision that it possessed exclusive jurisdiction. The plaintiffs in the district court cases could seek to revive their suits, leading to simultaneous challenges to the Clean Water Rule in multiple district courts. (A Judicial Panel on Multidistrict Litigation already rejected a request to consolidate the district court cases in 2015.) Further, each district court case will have its own appeals process, creating the potential for later disagreements in the U.S. courts of appeals on the underlying legality of the Rule.

Regardless of which courts hear challenges to the Clean Water Rule, administrative actions by the Trump Administration are likely to impact the litigation. In February 2017, the President issued an executive order directing the Corps and EPA to review and rescind or revise the Clean Water Rule. In a proposed rule published in July 2017 (analyzed here), the agencies initiated what they described as the first step in a two-step process to (1) rescind the Clean Water Rule and (2) engage in a separate rulemaking process to develop a new rule that will define the jurisdictional reach of the Clean Water Act.

EPA and the Corps currently are evaluating over 680,000 comments on the step-one proposal. But once the first step is complete and the Clean Water Rule is rescinded formally, any pending challenges to the Rule will likely be dismissed as moot (as discussed in this Sidebar).

DOES *NAM* ALLOW THE CORPS AND EPA TO START IMPLEMENTING THE CLEAN WATER RULE?

Before the Supreme Court's *NAM* decision, the Sixth Circuit had granted a nationwide stay of the Clean Water Rule, effectively halting its implementation across the United States. Although *NAM* did not vacate the Sixth Circuit's order granting the stay, the *NAM* Court concluded that the Sixth Circuit lacked subject matter jurisdiction over its consolidated appellate-level cases, and it instructed the Sixth Circuit to dismiss those cases in full. Consequently, once the Sixth Circuit enters a dismissal order, its stay of the Clean Water Rule will no longer be in place.

Even before the Sixth Circuit's stay, however, the United States District Court for the District of North Dakota had granted a preliminary injunction blocking operation of the Rule. While that injunction potentially could become effective again if litigation resumes at the district court level, the District of North Dakota did not enjoin implementation of the Clean Water Rule on a nationwide basis. As a result, once litigation resumes in the district courts, the Clean Water Rule could be enjoined in some parts of the country but allowed to operate in others.

Anticipating complications associated with an expiration of the Sixth Circuit's nationwide stay, the Corps and EPA initiated another rulemaking action in November 2017. This proposed rule seeks to add an "applicability date" to the Clean Water Rule of two years from the date of the final agency action on the proposal. According to the Corps and EPA, adding an applicability date to the Clean Water Rule would maintain the status quo and prevent inconsistencies and uncertainties about the regulatory regime that could arise after the nationwide stay is terminated, but before the agencies have finalized regulations rescinding the Clean Water Rule. In essence, the November 2017 proposed rule would allow the agencies more time to undertake the two-step rescind and replace process without allowing the 2015 Clean Water Rule to take effect during the rulemaking process.

CONGRESS AND THE CLEAN WATER RULE

Several legislative proposals introduced in the 115th Congress could impact the future of the Clean Water Rule. An omnibus appropriations bill that passed in the House would authorize the Corps and EPA to withdraw the Rule "without regard to any provision of statute or regulation that establishes a requirement for such withdrawal." If enacted, this provision could permit the agencies to complete the "step one" process of rescinding the Clean Water Rule without following the notice and comment rulemaking requirements of the Administrative Procedure Act (discussed in this Report). Other bills introduced in the 115th Congress would directly repeal the Rule or replace it by amending the Clean Water Act. Some Members of the House and Senate also have introduced resolutions expressing the sense of their Chamber that the Clean Water Rule should be withdrawn or vacated.

In: Water Quality
Editors: Edward Saunders et al.
ISBN: 978-1-53614-809-1

Chapter 5

WATER INFRASTRUCTURE FINANCING: HISTORY OF EPA APPROPRIATIONS*

Jonathan L. Ramseur and Mary Tiemann

SUMMARY

The principal federal program to aid municipal wastewater treatment plant construction is authorized in the Clean Water Act (CWA). Established as a grant program in 1972, it now capitalizes state loan programs through the clean water state revolving loan fund (CWSRF) program. Since FY1972, appropriations have totaled $97 billion.

In 1996, Congress amended the Safe Drinking Water Act (SDWA, P.L. 104-182) to authorize a drinking water SRF program to help systems finance projects needed to comply with drinking water regulations and to protect public health. Since FY1997, appropriations for the drinking water state revolving loan fund (DWSRF) program have totaled $21 billion.

The U.S. Environmental Protection Agency (EPA) administers both SRF programs, which annually distribute funds to the states for implementation. Funding amounts are specified in the State and Tribal Assistance Grants (STAG) account of EPA annual appropriations acts.

* This is an edited, reformatted and augmented version of Congressional Research Service, Publication No. 96-647, dated August 1, 2018.

The combined appropriations for wastewater and drinking water infrastructure assistance have represented 25%-32% of total funds appropriated to EPA in recent years.

Prior to CWA amendments in 1987 (P.L. 100-4), Congress provided wastewater grant funding directly to municipalities. The federal share of project costs was generally 55%; state and local governments were responsible for the remaining 45%. The 1987 amendments replaced the traditional grant program with the SRF program. Local communities are now often responsible for 100% of project costs, rather than 45%, because they are required to repay loans to states. The greater financial burden of the act's loan program on some cities has caused some to seek continued grant funding.

This concern has been particularly evident in the appropriations process where, for a number of years, Congress reserved as much as 30% of funds in the STAG account for special purpose grants directed to specified communities. Most of the funded projects were not authorized in the CWA or SDWA. Many state water quality officials and state infrastructure financing officials objected to this practice, as did EPA, since it reduced the amount of funding for state SRF programs. Due to general opposition to congressional earmarking, this practice has not occurred since FY2011.

Although the CWSRF and DWSRF have largely functioned as loan programs, both allow the implementing state agency to provide "additional subsidization" under certain conditions. Since its amendments in 1996, the SDWA has *authorized* states to use up to 30% of their DWSRF capitalization grants to provide additional assistance, such as forgiveness of loan principal or negative interest rate loans, to help disadvantaged communities (as determined by the state). Congress amended the CWA in 2014, adding similar provisions to the CWSRF program. In addition, appropriations acts in recent years have *required* states to use minimum percentages of their allotted SRF grants to provide additional subsidization.

The final full-year appropriations for FY2018 were enacted as part of the Consolidated Appropriations Act, 2018, signed by President Trump on March 23, 2018 (P.L. 115-141). Compared to FY2017, funding for EPA water infrastructure programs increased by 24% in FY2018. The act provided $1.694 billion for the CWSRF and $1,163 million for the DWSRF program, increases of $300 million to both programs compared to FY2017. The act also provided $63 million for the WIFIA program, more than doubling the FY2017 appropriation.

The Trump Administration's FY2019 budget request proposes the same amounts for the CWSRF and DWSRF programs as the FY2017 appropriation. However, the request proposes a 45% reduction in the state categorical grants compared to FY2018 levels.

INTRODUCTION

The Clean Water Act (CWA)[1] authorizes the principal federal program to aid municipal wastewater treatment plant construction and related eligible activities. Congress established this program in the Federal Water Pollution Control Act Amendments of 1972 (P.L. 92-500) (although prior versions of the act had authorized less ambitious grants assistance since 1956). Title II of P.L. 92-500 authorized grants to states for wastewater treatment plant construction under a program administered by the Environmental Protection Agency (EPA). Federal funds were provided through annual appropriations under a state-by-state allocation formula contained in the act itself. States used their allotments to make grants to cities to build or upgrade wastewater treatment plants, supporting the overall objectives of the act: restoring and maintaining the chemical, physical, and biological integrity of the nation's waters. The federal share of project costs, originally 75% under P.L. 92-500, was reduced to 55% in 1981.

By the mid-1980s, there was considerable policy debate between Congress and the Administration over the future of the act's construction grants program and, in particular, the appropriate federal role in funding municipal water infrastructure projects. Through FY1984, Congress had appropriated nearly $41 billion under this program, representing the largest nonmilitary public works programs since the Interstate Highway System. The grants program was a target of budget cuts in the Reagan Administration, which sought to redirect budgetary priorities in part to sort out the appropriate roles of federal, state, and local governments in a number of domestic policy areas, including water pollution control. The Administration's rationale included several points:

---

[1] The official statutory name is the Federal Water Pollution Control Act, P.L. 92-500, as amended, codified at 33 U.S.C. §1251 *et seq.*

- The original intent of the program to address the backlog of sewage treatment needs had been virtually eliminated by the mid-1980s.
- Most remaining projects (such as small, rural systems) were believed to pose little environmental threat and were not appropriate federal responsibilities.
- State and local governments, in the Administration's view, were fully capable of running construction programs and have a clear responsibility to construct treatment capacity to meet environmental objectives that were primarily established by states.

Thus, the Reagan Administration sought a phase-out of the act's construction grants program by 1990. Many states and localities supported the idea of phasing out the grants program, since many were critical of what they viewed as burdensome rules and regulations that accompanied the federal grant money. However, they sought a longer transition and ample flexibility to set up long-term financing to promote state and local self-sufficiency.

Congress's response to this debate was contained in 1987 amendments to the act (P.L. 100-4, the Water Quality Act of 1987). It authorized $18 billion over nine years for sewage treatment plant construction, through a combination of the Title II grants program and a new State Water Pollution Control Revolving Funds program—hereinafter the clean water state revolving fund (CWSRF) program. Under the new program, in CWA Title VI, federal grants would be provided as seed money for state-administered loans to build sewage treatment plants and, eventually, other water quality projects. Cities, in turn, would repay loans to the state, enabling a phase-out of federal involvement while the state built up a source of capital for future investments. Under the amendments, the CWSRF program was phased in beginning in FY1989 (in FY1989 and FY1990, appropriations were split equally between Title II and Title VI grants) and entirely replaced the previous Title II program in FY1991. The intention was that states would have flexibility to set priorities and administer funding, while federal aid would end after FY1994.

The CWSRF authorizations for appropriations provided in the 1987 amendments expired in FY1994, but pressure to extend federal funding has continued, in part because, although Congress has appropriated $97 billion in CWA Title II and Title VI wastewater infrastructure assistance since 1972, funding needs remain high: According to the most recent formal estimate by EPA and states, an additional $271 billion nationwide is needed over the next 20 years for all types of projects eligible for funding under the act.[2] Thus, Congress has continued to appropriate funds, and continued to assist states and localities in meeting wastewater infrastructure needs and comply with CWA requirements.

In 1996, Congress established a parallel program under the Safe Drinking Water Act (SDWA) to help communities finance projects needed to comply with federal drinking water regulations.[3] Funding support for drinking water occurred for several reasons. First, until the 1980s, the number of drinking water regulations was fairly small, and public water systems often did not need to make large investments in treatment technologies to meet those regulations. Second, good quality drinking water traditionally has been available to many communities at relatively low cost. By comparison, essentially all communities have had to construct or upgrade sewage treatment facilities to meet the requirements of the CWA.

Over time, drinking water circumstances changed, as communities grew, and commercial, industrial, agricultural, and residential land-uses became more concentrated, thus resulting in more contaminants reaching drinking water sources. Moreover, as the number of federal drinking water

---

[2] U.S. Environmental Protection Agency, *Clean Watersheds Needs Survey 2012, Report to Congress*, 2016, https://www.epa.gov/sites/production/files/2015-12/documents/cwns_2012_report_to_congress-508-opt.pdf.

[3] For information, see CRS Report RS22037, *Drinking Water State Revolving Fund (DWSRF): Program Overview and Issues*, by Mary Tiemann.

standards has increased, many communities have found that their water may not be as good as once thought and that additional treatment technologies are required to meet the new standards and protect public health. Between 1986 and 1996, for example, the number of regulated drinking water contaminants grew from 23 to 83, and EPA and the states expressed concern that many of the nation's 52,000 small community water systems were likely to lack the financial capacity to meet the rising costs of SDWA compliance.

Congress responded to these concerns by enacting the 1996 SDWA Amendments (P.L. 104-182), which authorized a drinking water state revolving loan fund (DWSRF) program to help systems finance projects needed to comply with SDWA regulations and to protect public health. This program, fashioned after the clean water SRF program, authorizes EPA to make grants to states to capitalize DWSRFs which states then use to make loans to public water systems. Appropriations for the program were authorized at $599 million for FY1994 and $1 billion annually for FY1995 through FY2003.

Capitalization grants for DWSRF programs were provided for the first time in FY1997. Although the authorizations for appropriations expired in FY2003, Congress has continued to provide funding for the program in annual appropriations, totaling $21 billion through FY2018. According to the most recent EPA-state survey, future funding needs for projects to treat and deliver public drinking water supplies in the United States are $473 billion over the next 20 years.[4]

The first section of this report includes a table that summarizes the history of appropriations for both wastewater and drinking water infrastructure programs. The next section discusses several historical developments in water infrastructure funding. The last section contains a detailed chronology of congressional activity regarding wastewater and drinking water infrastructure funding for each fiscal year since the 1987 CWA amendments.

---

[4] EPA, *Drinking Water Infrastructure Needs Survey and Assessment*, EPA-816-K-17-002, March 2018.

SUMMARY OF WATER INFRASTRUCTURE APPROPRIATIONS

Table 1 summarizes funding for the wastewater and drinking infrastructure programs since enactment of the 1987 CWA amendments (P.L. 100-4). Funding for these EPA programs is contained in the appropriations act providing funds for the Department of the Interior, Environment, and Related Agencies.[5] Within the portion of the bill that funds EPA, wastewater treatment assistance was first specified in an account called Construction Grants, which was subsequently renamed State Revolving Funds/Construction Grants, and then renamed Water Infrastructure. Since FY1996, this account has been titled State and Tribal Assistance Grants (STAG).

The STAG account now includes all water infrastructure funds and management grants provided to assist states in implementing air quality, water quality, and other media-specific environmental programs. The FY1996 appropriation was the first to include both water infrastructure and other state environmental grants; the latter previously were included in EPA's general program management account. Amounts shown in *Table 1* include funds for CWA Title II grants, CWSRF grants, drinking water SRF grants, special project grants (discussed below), and the Water Infrastructure Finance and Innovation Act (WIFIA) program.[6] Congress first provided appropriations to cover the subsidy costs of this program in FY2017, as discussed in the detailed chronology section below.

---

[5] Prior to the 109th Congress, EPA appropriations were included in acts funding the Department of Veterans Affairs, Department of Housing and Urban Development, and Independent Agencies (VA/HUD). In January 2005, House and Senate Appropriations Committees reorganized, and Congress moved jurisdiction over funding for EPA and several other entities to the Appropriations subcommittees covering Interior and Related Agencies.

[6] For more information, see CRS Report R43315, *Water Infrastructure Financing: The Water Infrastructure Finance and Innovation Act (WIFIA) Program*, by Jonathan L. Ramseur and Mary Tiemann.

Table 1. EPA Water Infrastructure Funding Millions of Dollars—Not Adjusted for Inflation

Fiscal Year	CWA Authorization	SDWA Authorization	President's Request	CWA Title II	CWSRF	DWSRF	Special Projects	WIFIA	Total Appropriation
1986	2,400		2,400	1,800					1,800
1987	2,400		2,000	2,361					2,361
1988	2,400		2,000	2,304					2,304
1989	2,400		1,500	941	941		68		1,950
1990	2,400		1,200	960	967		53		1,980
1991	2,400		1,600		2,048		36		2,084
1992	1,800		1,883		1,949		435		2,384
1993	1,200		2,467		1,928		556		2,484
1994	600	599	2,047		1,218		558		1,776
1995	—	1,000	2,528		1,235		834		2,069
1996	—	1,000	2,365		2,074		307		2,380
1997	—	1,000	2,178		625	1,275	301		2,201
1998	—	1,000	2,078		1,350	725	393		2,468
1999	—	1,000	2,028		1,350	775	402		2,527
2000	—	1,000	1,753		1,345	820	395		2,561
2001	—	1,000	1,753		1,350	825	466		2,641
2002	—	1,000	2,233		1,350	850	459		2,659
2003	—	1,000	2,185		1,341	845	413		2,599
2004	—	—	1,798		1,342	845	425		2,612
2005	—	—	1,794		1,091	843	402		2,336
2006	—	—	1,649		887	838	281		2,005

Fiscal Year	CWA Authorization	SDWA Authorization	President's Request	CWA Title II	CWSRF	DWSRF	Special Projects	WIFIA	Total Appropriation
2007	—	—	1,570		1,084	838	84		2,005
2008	—	—	1,553		689	829	177		1,695
2009[a]	—	—	1,397		4,689	2,829	184		7,702
2010	—	—	3,920		2,100	1,387	187		3,674
2011	—	—	3,307		1,522	963	20		2,505
2012	—	—	2,560		1,467	918	15		2,399
2013[b]	—	—	2,045		1,376	861	14		2,252
2014	—	—	1,927		1,449	907	15		2,371
2015	—	—	1,790		1,449	907	15		2,371
2016	—	—	2,317		1,394	863	30		2,287
2017	—	—	2,022		1,394	963	30	33	2,420
2018[c]	—	—	2,257		1,694	1,163	30	63	2,950
2019			2,280		1,394 (R)	863 (R)	3 (R)	20 (R)	

Source: Compiled by CRS from annual appropriations acts. Notes: (R) = requested.

[a] The FY2009 total includes $6.0 billion in supplemental appropriations provided under the American Recovery and Reinvestment Act, P.L. 111-5, consisting of $4.0 billion for CWA SRF capitalization grants and $2.0 billion for SDWA SRF capitalization grants.

[b] FY2013 total reflects post-sequester/post-rescission amounts. See text for detail.

[c] The FY2018 appropriation provided new funding (in aggregate $50 million) for three programs authorized in the Water Infrastructure Improvements for the Nation (WIIN) Act (P.L. 114-322, Title II, the Water and Waste Act of 2016. This funding is not included in the table.

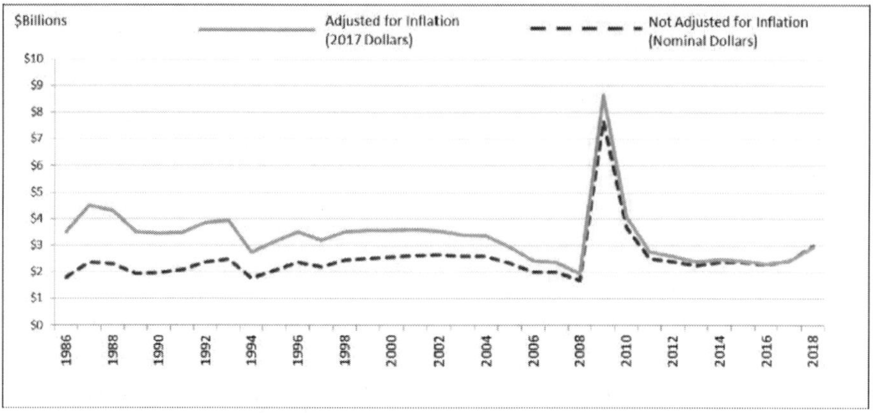

Source: Prepared by CRS using information from annual appropriations acts, committee reports, and explanatory statements presented in the *Congressional Record.* Amounts reflect applicable rescissions and supplemental appropriations, including $6 billion in the American Recovery and Reinvestment Act of 2009 (P.L. 111-5). Constant dollars calculated from Office of Management of Budget, Table 10.1, "Gross Domestic Product and Deflators Used in the Historical Tables: 1940-2023," https://www.whitehouse.gov/omb/budget/Historicals.

Figure 1. EPA Water Infrastructure Annual Appropriations: FY1986-FY2018; Adjusted ($2017) and Not Adjusted for Inflation (Nominal).

Table 1 does not include funds for consolidated state environmental management grants. These grants include funding for a wide range of environmental programs, which have changed over time. In recent years, the categorical grants have included funding for water, air, and waste programs. The categorical grant programs most closely related to water infrastructure issues include grants for states' nonpoint source management programs (CWA Section 319) and states' pollution control programs (CWA Section 106). Funding levels for the environmental management state grants are discussed below in the appropriations chronology section.

As an additional comparison, Figure 1 illustrates the total EPA water infrastructure appropriations (for clean water and drinking water assistance combined) between FY1986 and FY2018 in both nominal dollars (i.e., not adjusted for inflation) and constant (2017) dollars (i.e., adjusted for inflation).

HISTORICAL FUNDING DEVELOPMENTS

This section discusses several historical developments of note regarding water infrastructure appropriations.

Special Purpose Project Grants

The practice of earmarking a portion of the construction grants/SRF account for specific wastewater treatment and other water quality projects began with the FY1989 appropriations. The practice increased to the point of representing a significant portion of appropriated funds (31% of the total water infrastructure appropriation in FY1994, for example, but less in subsequent years: 2.5% in FY2009 and 5% in FY2010). The number of projects receiving these earmarked funds also increased: from 4 in FY1989 to 319 in FY2010. Beginning in FY2000, the larger total number of earmarked projects resulted in more communities receiving such grants, but at the same time receiving smaller amounts of funds. Thus, while a few communities received individual earmarked awards of $1 million or more, the average size of earmarked grants shrank: $18.1 million in FY1995, $4.9 million in FY1999, $1.08 million in FY2006, and $586,000 in FY2010. (Conference reports on the individual appropriations bills, noted in the later discussion in this report, provide some detail on projects funded in this manner.) The effective result of earmarking was to reduce the amount of funds provided to states to capitalize their SRF programs. Between FY1989 and FY2010, approximately 10% of the total water infrastructure appropriations ($7.4 billion) went to earmarked project grants.[7]

Interest groups representing state water quality program managers and administrators of infrastructure financing programs criticized the practice of earmarked appropriations. They contended that earmarking undermined

---

[7] For additional information, see CRS Report RL32201, *Water Infrastructure Projects Designated in EPA Appropriations: Trends and Policy Implications*, by Claudia Copeland.

the intended purpose of the state funds—promoting water quality improvements nationwide. Many state officials preferred funds to be allocated more equitably, not based on what they viewed largely as political considerations, and they preferred for state environmental and financing officials to retain responsibility to set actual spending priorities. Further, they argued that the special projects funding would diminish the level of seed funding to SRFs, delaying the time when SRFs would be financially self-sufficient.

The practice of earmarking was criticized because designated projects were arguably receiving more favorable treatment than other communities' projects: They were generally eligible for 55% federal grants (and were not required to repay 100% of the funded project cost, as is the case with a loan through an SRF), and the practice circumvented the standard process of states determining the priority by which projects will receive funding. It also meant that the projects were generally not reviewed by the CWA authorizing committees. This was especially true after FY1992, when special purpose grant funding was designated for types of projects not authorized in the Clean Water Act or the Safe Drinking Water Act.

Members of Congress intervened for a specific community for a number of reasons. In some cases, the communities may have been unsuccessful in seeking state approval to fund the project under an SRF loan or other program. For some, the cost of a project financed through a state loan was deemed unacceptably high, because repaying the loan would result in increased user fees that ratepayers felt would have been unduly burdensome.

In the early years of this congressional practice, special purpose grant funding originated in the House version of the EPA appropriations bill, while the Senate, for the most part, resisted earmarking by rejecting or reducing amounts and projects included in House-passed legislation. Therefore, special purpose grant funding on several occasions was an issue during the House-Senate conference on the appropriations bill. Beginning in FY1999, however, both the House and Senate proposed earmarked projects in their respective versions of the EPA appropriations bill, with the final total number of projects and dollar amounts determined by conferees.

The Clean Water Act Title II grants program effectively ended when authorizations for it expired after FY1990. One result of earmarking special purpose grants in appropriations bills was to perpetuate grants as a method of funding wastewater treatment construction long after FY1990. This practice led Congress to provide EPA grants for drinking water system projects, which had not previously been available. However, as discussed in the next section, general opposition to congressional earmarking stopped the practice after FY2011.

Local Cost Share on Special Purpose Grants

The federal percentage share and local match required on special purpose grants varied depending on the project and the year of funding. For example, in the early projects (FY1989), the 1987 CWA amendments specified the federal cost shares, which ranged from 75% to 85%. In FY1992 and FY1993, the appropriations acts specified that funds were provided "as grants under title II," resulting in a requirement for local communities to provide a 45% share of project costs. After FY1993, the appropriations acts themselves were the authority for the special purpose projects grants. In the FY1995 appropriation bill, which also directed allocation of funds appropriated in FY1994 to several needy cities, Congress addressed the issue of federal and local cost shares in report language accompanying the bill, but not in the appropriation act itself.[8]

> The conferees are in agreement that the agency should work with the grant recipients on appropriate cost-share arrangements. It is the conferees' expectation that the agency will apply the 45% local cost share requirement under Title II of the Clean Water Act in most cases.

---

[8] H.Rept. 103-715, accompanying H.R. 4624, 103rd Cong., 2d sess., p. 42.

In the FY1996 appropriations, both the act and accompanying reports were silent on federal/local cost share and applicability of Title II requirements. Because of that, EPA officials planned to require only a 5% local match for most of the special purpose grants in that bill, which is the standard matching requirement for other EPA noninfrastructure grants. Under the agency's rules, the local match could include in-kind services, as well as funding toward the project.

In the FY1997 appropriations, Congress included report language as it had in FY1995 concerning federal and local cost share requirements.[9]

> The conferees are in agreement that the Agency should work with the grant recipients on appropriate cost-share agreements and to that end the conferees direct the Agency to develop a standard cost-share consistent with fiscal year 1995.

The FY1998 and FY1999 appropriations included neither bill nor report language on this point. However, language in the House and Senate Appropriations Committees' reports on the FY1998 and FY1999 bills directed EPA to work with grant recipients on appropriate cost-share arrangements.[10]

For FY2000, Congress included explicit report language concerning the local match.[11]

> The conferees agree that the $331,650,000 provided to communities or other entities for construction of water and wastewater treatment facilities and for groundwater protection infrastructure shall be accompanied by a cost-share requirement whereby 45 percent of a project's cost is to be the responsibility of the community or entity consistent with longstanding guidelines for the Agency. These guidelines also offer flexibility in the application of the cost-share requirement for those few circumstances when meeting the 45 percent requirement is not possible.

---

[9] H.Rept. 104-812, accompanying H.R. 3666, 104th Cong., 1st sess., p. 74.

[10] H.Rept. 105-175, accompanying H.R. 2158, 105th Cong., 1st sess., p. 69; S.Rept. 105-216, accompanying S. 2168, 105th Cong., 2nd sess., p. 82.

[11] H.Rept. 106-379, accompanying H.R. 2684, 106th Cong., 1st sess., p. 141.

Similar report language concerning local cost-share requirements accompanied the conference reports on the appropriations bills from FY2001 through FY2005. Beginning with FY2004, Congress specified in the appropriations legislation that the local share of project costs shall be not less than 45%. Similarly, beginning with the FY2003 appropriations legislation, Congress also specified that, except for those limited instances in which an applicant meets the criteria for a waiver of the cost-share requirement, the earmarked grant shall provide no more than 55% of an individual project's cost, regardless of the amount appropriated.

The practice of earmarking special project water infrastructure grants continued to change. First, in FY2007, Congress applied a one-year moratorium on earmarks in all appropriations bills. For the next three years, special project grants were allowed in appropriations bills—including EPA's—but again in FY2011, no special project funding was provided for congressional projects. Following the 2010 mid-term election and during subsequent months while FY2011 appropriations were under consideration (discussed below), the general issue of congressional earmarks of specific projects had become highly controversial because of the overall growing number of them, concern over the influence of special interests on spending decisions, and lack of congressional oversight. In response, President Obama said he would veto any legislation containing earmarks, the House extended the ban on earmarks under the Republican Conferences rules, and the chairman of the Senate Appropriations Committee announced a moratorium on earmarks for FY2011 and FY2012. Thus, the FY2011 full-year appropriations measure contained no congressionally directed special project funds for water infrastructure projects in the EPA STAG account. However, it did include funds requested by the President: $10 million for Alaska Native Villages and $10 million for U.S.-Mexico border projects.[12]

The FY2012 full-year appropriations measure also contained no special project funding in the EPA STAG account. The FY2012 bill did

---

[12] These appropriations for these purposes had been authorized in P.L. 104-182, the Safe Drinking Water Act Amendments of 1996.

include funds for Alaska Native and Rural Villages ($10 million) and for U.S.-Mexico border projects ($5 million).

The moratorium on congressional earmarks has continued. The FY2013 full-year appropriations measure (P.L. 113-6) contained no special project funding in the STAG account. As with other recent bills, however, it did include funds for Alaska Native and Rural Villages ($9.5 million) and for U.S.-Mexico border projects ($4.7 million). Similarly, the moratorium on earmarks continued in FY2014 and FY2015; P.L. 113-76 contained no special project funding in the STAG account for FY2014, but did include funds for Alaska Native and Rural Villages ($10 million) and for U.S.-Mexico border projects ($5 million). The FY2015 funding bill, P.L. 113-235, was the same as FY2014. Both the FY2016 and FY2017 appropriations acts (P.L. 114-113 and P.L. 115-31, respectively) included $10 million for Alaska Native and Rural Villages and $20 million for U.S. Mexico border projects. President Trump's FY2018 budget request proposes to eliminate funding for these two programs.

Additional Subsidization

Although the CWSRF and DWSRF have largely functioned as loan programs, both allow the implementing state agency to provide "additional subsidization" under certain conditions. Since its amendments in 1996, the SDWA has authorized states to use up to 30% of their DWSRF capitalization grants to provide additional assistance, such as forgiveness of loan principal or negative interest rate loans, to help disadvantaged communities (as determined by the state).[13] Congress amended the CWA in 2014, adding similar provisions to the CWSRF program.[14]

In addition, appropriations acts in recent years have *required* states to use minimum percentages of their allotted funds to provide additional subsidization. This trend began with the American Recovery and Reinvestment Act of 2009 (P.L. 111-5), which required states to use at

[13] 42 U.S.C. §300j-12(d).
[14] 33 U.S.C. §1383(i).

least 50% of their funds to "provide additional subsidization to eligible recipients in the form of forgiveness of principal, negative interest loans or grants or any combination of these." Subsequent appropriation acts have included similar conditions, with varying percentages of subsidization. The FY2016, FY2017, and FY2018 appropriations acts included an identical condition, requiring 10% of the CWSRF grants and 20% of the DWSRF grants to be used "to provide additional subsidy to eligible recipients in the form of forgiveness of principal, negative interest loans, or grants (or any combination of these)."

Non-Infrastructure Grants

The 1987 CWA amendments authorized federal grants to assist states in implementing programs to manage water pollution from nonpoint sources such as farm and urban areas, construction, forestry, and mining sites. Because of competing demands for funding, it was difficult for Congress to fund this grant program and other water quality initiatives in the 1987 act. Appropriators did fund Section 319 grants in EPA's general program management account (abatement, control, and compliance) in FY1990, FY1991, and FY1992 but well below authorized levels. In the FY1993 act, appropriators moved funding into the SRF/construction grants account, thereby providing a degree of protection from competing priorities. In FY1996, Congress included all state grants for management of environmental programs in a single consolidated grants appropriation. In doing so, Congress endorsed a Clinton Administration proposal for a more flexible approach to state grants, a key element of EPA's efforts to improve the federal-state partnership in environmental programs. In more recent years, Congress has provided specific funding amounts for certain programs within the categorical grants appropriation.

APPROPRIATIONS CHRONOLOGY

This section summarizes, in chronological order, congressional activity to fund items in the STAG account since the 1987 CWA amendments.

FY1986, FY1987

The authorization period covered by P.L. 100-4 was FY1986-FY1994. By the time the amendments were enacted, FY1986 was over, as was a portion of FY1987. Thus, appropriations for those two years only indirectly reflected the policy and program changes for later years that were contained in P.L. 100-4. For FY1986, Congress appropriated a total of $1.8 billion, consisting of $600 million approved in December 1985 (while Congress was beginning to debate reauthorization legislation that eventually was enacted as P.L. 100-4 in January 1987) and $1.2 billion more in July 1986.

For FY1987, while debate on CWA reauthorization continued, President Reagan requested $2.0 billion, consistent with his legislative proposal to terminate the grants program by FY1990. In October 1986, Congress appropriated $2.4 billion (P.L. 99-500/P.L. 99-591). However, only $1.2 billion of that amount was released immediately, pending enactment of a reauthorization bill, which was then in conference.

Following enactment of the Water Quality Act of 1987, remaining FY1987 funds were released as part of a supplemental appropriations bill (P.L. 100-71). Conferees on that measure agreed, however, to shift $39 million of the remaining unreleased grant funds to other priority water quality activities authorized in P.L. 100-4. The final total of construction grant monies was $2.361 billion.

FY1988

For FY1988 the President again requested $2.0 billion. In December 1987, Congress approved legislation providing FY1988 appropriations (P.L. 100-202, the omnibus continuing resolution to fund EPA and other federal agencies). In it, Congress appropriated $2.304 billion for construction grants. Final action on the EPA budget and other funding bills had been delayed by budget-cutting talks between Congress and the White House. Reduced construction grants funding was one of many spending cuts required to implement a congressional-White House "summit agreement" on the budget. The final construction grants appropriation was less than funding levels that had been included in separate versions of a bill passed by the House and Senate before the budget summit, $2.4 billion.

FY1989

For FY1989, President Reagan requested $1.5 billion, or 35% below FY1988 appropriations and 37.5% less than the authorized level of $2.4 billion for FY1989. In separate versions of an EPA appropriations bill, the House and Senate voted to provide $1.95 billion and $2.1 billion respectively. The final figure, in P.L. 100-404, was $1.95 billion, which included $68 million for special projects in four states. Thus, the actual amount provided for grants was $1.882 billion. That total was divided equally between the previous Title II grants program and new Title VI SRF program, as provided in the authorizing language of P.L. 100-4.

The FY1989 legislation was the first to include earmarking of funds for specified projects or grants in EPA's construction grants account, an action that continued in subsequent years, as discussed above. All of the projects funded in the 1989 legislation were ones that had been authorized in provisions of the Water Quality Act of 1987 (WQA, P.L. 100-4). The designated projects were in Boston (authorized in Section 513 of the WQA, to fund the Boston Harbor wastewater treatment project), San Diego/Tijuana (Section 510, to fund an international sewage treatment project needed because of the flow of raw sewage from Tijuana, Mexico, across the border), Des Moines, IA (Section 515, for sewage treatment plant construction), and Oakwood Beach/Redhook, NY (Section 512 of the WQA, to relocate natural gas distribution facilities that were near wastewater treatment works in New York City).

FY1990

For FY1990, President Reagan's budget requested $1.2 billion in wastewater treatment assistance, or 50% less than the authorized level and 38.5% less than the FY1989 enacted amount of $1.95 billion. Further, the Reagan budget proposed that the $1.2 billion consist of $800 million in Title VI monies and $400 million in Title II grants, contrary to provisions of the CWA directing that appropriations be equally divided between the two grant programs, as in FY1989. President Bush's revised FY1990 budget, presented in March 1989, made no changes from the Reagan budget in this area.

In acting on this request, Congress agreed to provide $2.05 billion, including $46 million for three special projects (Boston, San Diego/Tijuana, and Des Moines), leaving a total of $1.002 billion each for Titles II and VI (P.L. 101-144). Title II funds were reduced by $6.8 million, however, due to funds earmarked for a specific project in South Carolina. Although these amounts were appropriated, all funds in the bill were reduced by 1.55% (or, a $31.8 million reduction from the construction grants account) to provide funds for the federal government's antidrug program.

Final FY1990 appropriations were altered again by passage of the FY1990 Budget Reconciliation measure and implementation of the Balanced Budget and Emergency Deficit Control Act (the Gramm-Rudman-Hollings Act), which established procedures to reduce budget deficits annually, resulting in a zero deficit by 1993. For each fiscal year that the deficit was estimated to exceed maximum targets established in law, an automatic spending reduction procedure was triggered to eliminate deficits in excess of the targets through "sequestration," or permanent cancellation of budgetary resources.

Thus, to meet budget reduction mandates and, in particular, deficit reduction targets under the Balanced Budget and Emergency Deficit Control Act (the Gramm-Rudman-Hollings Act), additional funding cuts were included in P.L. 101-239, the Budget Reconciliation Act of 1989, affecting construction grants funding and all other accounts not exempted from Gramm-Rudman procedures. P.L. 101-239 provided that the "sequestration" procedures under the GrammRudman-Hollings Act would be allowed to apply for a portion of FY1990 (for 130 days, or 35.6% of the year), providing an additional automatic spending reduction in EPA and other agencies' programs subject to the act.

As a result of these reductions, funding for wastewater treatment aid in FY1990 totaled $1.98 billion, or $30 million more than in FY1989. The total included $53 million for special projects in San Diego, Boston, Des Moines, and Honea Path/Ware Shoals, SC, $960 million for Title II grants, and $967 million for Title VI grants. The combined reductions amounted to 3.4% less than the amount agreed to by conferees on P.L. 101-144 (i.e., $2.05 billion), before subtracting funds for antidrug programs and accounting for effects of the Gramm-Rudman partial-year sequester.

FY1991

For FY1991, President Bush requested $1.6 billion in funding for wastewater treatment assistance. This total included $15.4 million to be earmarked for the San Diego project authorized in Section 510 of the Water Quality Act of 1987, to fund construction of an international sewage treatment project. The remainder, $1.584 billion, would be only for capitalization grants under Title VI of the act, as the 1987 legislation provides for no new Title II grants after FY1990.

In acting on EPA's appropriations for FY1991 (P.L. 101-507), Congress agreed to provide $2.1 billion in wastewater treatment assistance. Beginning in FY1991, all appropriated funds are utilized for capitalization grants under Title VI of the act (as provided in the Water Quality Act of 1987); funding for the traditional Title II grants program was no longer available.

The enacted level included several earmarkings: $15.7 million for San Diego, $20 million for Boston Harbor (Section 513 of the WQA), and $16.5 million for a new Water Quality Cooperative Agreement Program under Section 104(b)(3) of the act.[15] The President's budget had requested $16.5 million to support state permitting, enforcement, and water quality management activities, especially to offset the reductions in aid to states due to elimination of state management setasides from the previous Title II construction grants program. Congress agreed to the level requested, but provided it as a portion of the wastewater treatment appropriation, rather than as part of EPA's general program management appropriation, as in the President's request. As a result of these earmarkings, $2.048 billion was provided for Title VI grants.

FY1992

For FY1992, President Bush requested $1.9 billion in wastewater treatment funds, or $100 million more than authorized under the Water Quality Act of 1987 for Title VI grants in FY1992. However, out of the $1.9 billion total, the President's request sought $1.5 billion for Title VI SRF grants and $400 million as grants under the expired Title II construction grants program for the following coastal cities: Boston, San Diego, New York, Los Angeles, and Seattle. Two of the five designated projects had been authorized in the 1987 CWA amendments; the other three did not have explicit statutory authorization. Also, $16.5 million was requested for Water Quality Cooperative Agreement grants to the states.

---

[15] Section 104(b)(3) grants have been used to support a variety of special studies and projects allowing states and localities to demonstrate innovative approaches to implementing the core water quality program.

In acting on the request in November 1991, Congress provided total wastewater funds of $2.4 billion (P.L. 102-139). The total was allocated as follows:

- $1,948.5 million for SRF capitalization grants,
- $16.5 million for Section 104(b)(3) grants,
- $49 million for the special project in San Diego-Tijuana (Section 510 of the Water Quality Act),
- $46 million to the Rouge River (MI) National Wet Weather Demonstration Project, and
- $340 million as construction grants under title II of the Clean Water Act for several other special projects—the Back River Wastewater Treatment Plant (Baltimore), Maryland, the Boston Harbor project, New York City, Los Angeles, San Diego (a wastewater reclamation project), and Seattle.

This appropriation bill was the first to include special purpose grant funding for several projects not specifically authorized in the Clean Water Act or amendments to that law.

FY1993

For FY1993, President Bush requested $2.484 billion for state revolving funds/construction grants (now called the water infrastructure account). The requested total included $340 million to be targeted for 55% construction grants to six communities: Boston, New York, Los Angeles, San Diego, Seattle, and Baltimore. In addition, the President requested that $130 million be directed toward a Mexican Border Initiative, consisting of $65 million for construction of the international treatment plant at San Diego (to address the Tijuana sewage problem), $15 million for projects at Nogales, AZ, and New River, CA, and $50 million as 50% grants for *colonias* in Texas.[16] The President also requested $16.5 million for Section 104(b)(3) grants. Along with these special project and grant amounts, the request sought $2.014 billion for SRF assistance.

Final action on FY1993 funding occurred on September 25, 1992 (P.L. 102-389). It provided an appropriation of $2.55 billion, but $622.5 million of this amount was reserved for special projects and other grants. The bill provided $50 million in CWA Section 319 grants and $16.5 million in Section 104(b)(3) grants out of the SRF amount. It included $556 million for the following special purpose grants: the international treatment plant at San Diego (Tijuana—Section 510 of the WQA, with bill language capping funding for that project at $239.4 million), plus projects in Boston; New York; Los Angeles; San Diego; Seattle; Rouge River, MI; Baltimore; Ocean County, NJ; Atlanta; and for *colonias* in Texas, Arizona, and New Mexico. The final SRF grant amount under the bill was $1.928 billion.

Early in 1993, President Clinton requested that Congress approve "economic stimulus and investment" spending, in the form of supplemental FY1993 appropriations. Both his original proposal and a subsequent modified proposal included additional SRF grant funds, but neither of the bills enacted by Congress in response to these requests (P.L. 103-24, P.L. 103-50) provided additional SRF funds.

FY1994

For FY1994, the Clinton Administration requested $2.047 billion for water infrastructure. The funds in this request were $1.198 billion to capitalize State Revolving Funds, $150 million for Mexican border project grants, and $100 million for a single hardship community (Boston). The request also included $599 million to capitalize new state drinking water revolving funds.

The final version of the FY1994 legislation (P.L. 103-124) provided $2.477 billion for water infrastructure/state revolving funds. Of this total

[16] *Colonias* are unincorporated areas outside city boundaries along the U.S.-Mexico border. Most lack adequate public utilities, especially water and wastewater services.

amount, $599 million was to be reserved for drinking water SRFs, if authorization legislation were enacted; $80 million was for Section 319 grants; $22 million was for Section 104(b)(3) grants; and $58 million was for Tijuana/San Diego—Section 510 of the WQA. This resulted in an appropriation of $1.718 billion for clean water SRFs.

In addition, the final bill provided that $500 million be used to support water infrastructure financing in economically distressed/hardship communities. Under the bill, these funds were not available for spending until May 31, 1994, and were set aside until projects were authorized in the CWA for this purpose.

Thus, the bill as enacted provided $1.218 billion immediately for clean water SRFs, with the expectation that $500 million more would be available for financing hardship community projects after May 31, 1994.

FY1995

For FY1995, President Clinton requested $2.65 billion for water infrastructure consisting of $1.6 billion for CWA SRFs, $100 million for Section 319 nonpoint source management grants to states, $52.5 million for a grant to San Diego for a wastewater project pursuant to Section 510 of the WQA, $47.5 million for other Mexican border projects, $50 million to the state of Texas for *colonias* projects, and $100 million for grants under Title II for needy cities (intended for Boston). The request included $700 million for drinking water SRFs, pending enactment of authorizing legislation. The President's budget also requested $21.5 million for Section 104(b)(3) grants/cooperative agreements.

Final agreement on FY1995 funding was contained in P.L. 103-327, enacted in September 1994, which provided a total of $2.962 billion for water infrastructure financing. Of the total, $22.5 million was for grants under Section 104(b), $100 million for Section 319 grants, $70 million for Public Water System Supervision program grants (grants to states under the Safe Drinking Water Act to support state implementation of delegated drinking water programs), $52.5 million for the Section 510 project in San

Diego, and $700 million for drinking water SRFs (contingent upon enactment of authorization legislation).

The remaining $2.017 billion was for CWA projects. Of this amount, $1.235 billion was for clean water SRF grants to states under Title VI of the CWA. The remaining $781.8 million (39% of this amount, 26% of the total appropriation) was designated for 45 specific, named projects in 22 states. The earmarked amounts ranged in size from $200,000 for Southern Fulton County, PA, to $100 million for the city of Boston.

Finally, the conferees included bill language concerning release of the $500 million in FY1994 needy cities money (because the authorizing committees of Congress had not acted on legislation to authorize specific projects, as had been intended in P.L. 103-124) as follows:

- $150 million to Boston, $50 million for *colonias* in Texas, $10 million for *colonias* in New Mexico, $70 million for a New York City wastewater reclamation facility, $85 million for the Rouge River project, $50 million for the city of Los Angeles, $50 million for the county of Los Angeles, and $35 million for Seattle, WA.

FY1996

In February 1995, President Clinton submitted the Administration's budget request for FY1996. It requested $2.365 billion for water infrastructure funding consisting of $1.6 billion for clean water state revolving funds, $500 million for drinking water state revolving funds, $150 million to support Mexico border projects under the U.S.-Mexican Border Environmental Initiative and NAFTA, and $100 million for special need/economically distressed communities (not specified in the request, but presumed to be intended for Boston), plus $15 million for water infrastructure needs in Alaska Native Villages.

In February 1995, congressional appropriations committees began considering legislation to rescind previously appropriated FY1995 funds, as part of overall efforts by the 104th Congress to shape the budget and federal spending. These efforts resulted in passage in July 1995 of P.L. 104- 19, which rescinded $16.5 billion in total funds from a number of departments, agencies, and programs. In the water infrastructure area, it rescinded $1,077,200,000 from prior year appropriations including the $3.2 million for a project in New Jersey (it had mistakenly been funded twice in P.L. 103-327) and $1,074,000,000 in other water infrastructure appropriations. Although not contained in bill language, it was understood that the larger rescinded amount consisted solely of drinking water SRF funds (leaving $1.235 billion for FY1995 clean water SRF funds, $778.6 million for earmarked wastewater projects—both amounts as originally appropriated—and $225 million in FY1994-FY1995 drinking water SRF funds that had not yet been authorized).

It took until April 1996 for Congress and the Administration to reach agreement on FY1996 appropriations for EPA as part of omnibus legislation (P.L. 104-134) that consolidated five appropriations bills not yet enacted due to disagreements over funding levels and policy. Agreement came as the fiscal year was more than one-half over.

Before that, however, congressional conferees reached agreement in November 1995 on FY1996 legislation for EPA (H.R. 2099, H.Rept. 104-353). Conferees agreed to provide $2.323 billion for a new account titled State and Tribal Assistance Grants, consisting of infrastructure assistance and state environmental management grants for 16 categorical programs that had previously been funded in a separate appropriations account. The total included $1.125 billion for clean water SRF grants, $275 million in new appropriations for drinking water SRF grants, and $265 million for special purpose project grants. Report language provided that the drinking water SRF money also included $225 million from FY1995 appropriations rescinded in P.L. 104-19. The drinking water SRF money would be available upon enactment of SDWA reauthorization legislation that would authorize a drinking water SRF program; otherwise, it would revert to clean water SRF grants if the SDWA were not reauthorized by June 30, 1996. This made the total potentially available for drinking water SRF grants $500 million.

The November 1995 agreement on H.R. 2099 included $658 million for consolidated state environmental grants. In doing so, Congress endorsed an Administration proposal for a more flexible approach to state grants, a key element of EPA's efforts to improve the federal-state partnership in environmental programs. In lieu of traditional grants provided separately to support state air, water, hazardous waste, and other programs, consolidated grants are intended to reduce administrative burdens and improve environmental performance by allowing states and tribes to target funds to meet their specific needs and integrate their environmental programs, as appropriate. Congress's support was described in accompanying report language.[17]

> The conferees agree that Performance Partnership Grants are an important step to reducing the burden and increasing the flexibility that state and tribal governments need to manage and implement their environmental protection programs. This is an opportunity to use limited resources in the most effective manner, yet at the same time, produce the results-oriented environmental performance necessary to address the most pressing concerns while still achieving a clean environment.

Including state environmental grants in the same account with water infrastructure assistance reflected Congress's support for enhancing the ability of states and localities to implement environmental programs flexibly and support for EPA's ability to provide block grants to states and Indian tribes.

The H.R. 2099 conference agreement also included legislative riders intended to limit or prohibit EPA from spending money to implement several environmental programs. The Administration opposed the riders. The House and Senate approved this bill in December, but President Clinton vetoed it, because of objections to spending and policy aspects of the legislation.

---

[17] H.Rept. 104-384, accompanying H.R. 2099, 104th Cong., 1st sess., in Congressional Record, vol. 141, no. 193, daily ed. December 6, 1995, p. H 14132. This was the second conference report on this bill; a previous agreement, reflected in H.Rept. 104-353, was rejected by the House on November 29. However, amounts in the State and Tribal Assistance Grants account were the same in both versions.

With no full-year funding in place from October 1995 to April 1996, EPA and the programs it administers (along with agencies and departments covered by four other appropriations bills not yet enacted) were subject to a series of short-term continuing resolutions, some lasting only a day, some lasting several weeks. In March 1996, the House and Senate began consideration of an omnibus appropriations bill to fund EPA and other agencies for the remainder of FY1996, finally reaching agreement in April on a bill (H.R. 3019) enacted as P.L. 104-134.[18] Congress agreed to provide $2.813 billion for a new account titled State and Tribal Assistance Grants, consisting of state grants and infrastructure assistance, as in H.R. 2099, the vetoed measure. The total was divided as follows:

- $1.3485 billion for clean water SRF grants (including $50 million for impoverished communities),
- $500 million in new appropriations for drinking water SRF grants,
- $150 million for Mexico-border project grants and Texas *colonias*, as requested,
- $15 million for Alaska Native Villages, as requested,
- $141.5 million for 17 special purpose project grants, and
- $658 million for consolidated state environmental grants, which states could use to administer a range of delegated environmental programs.

Report language provided that the drinking water SRF money also included $225 million from FY1995 appropriations that remained available after the rescissions in P.L. 104-19, for a total of $725 million. The drinking water SRF money was contingent upon enactment of legislation authorizing an SRF program under the Safe Drinking Water Act by August 1, 1996; otherwise, it would revert to clean water SRF grants.

---

[18] The conference report on H.R. 3019 (H.Rept. 104-537) references the conference report on the vetoed H.R. 2099, making the two reports together the full statement of the conference committee regarding EPA funding and the State and Tribal Assistance Grants account.

The final agreement (P.L. 104-134) included several of the legislative riders from previous versions of the legislation, including riders related to drinking water and clean air, but dropped others strongly opposed by the Administration.

Funds within the State and Tribal Assistance Grants account were redistributed after Congress passed Safe Drinking Water Act amendments in August 1996. Enactment of the amendments (P.L. 104-182) occurred on August 6—after the August 1 deadline in P.L. 104-134 that would have made $725 million available for drinking water SRF grants in FY1996. Thus, the previously appropriated $725 million reverted to clean water SRF grants, making the FY1996 total for those grants $2.0735 billion.

FY1997

While debate over the FY1996 appropriations was continuing, in March 1996, President Clinton submitted the details of a FY1997 budget. For water infrastructure and state and tribal assistance, the request totaled $2.852 billion consisting of

- $1.35 billion for clean water SRF grants (the request included language that would authorize states the discretion to use this SRF money either for clean water or drinking water projects),
- $165 million for U.S.-Mexico border projects, Texas *colonias*, and Alaska Native Village projects,
- $113 million for needy cities projects,
- $550 million for drinking water infrastructure SRF funding, contingent upon enactment of authorizing legislation, and
- $674 million for state performance partnership consolidated management grants, which could address a range of environmental programs.

In response to the Administration's request, in June 1996 the House approved legislation (H.R. 3666) providing FY1997 funding for EPA. In the State and Tribal Assistance Grants account, the House approved $2.768 billion, $84 million less than requested but on the whole endorsing the budget request. The total provided the following: $1.35 billion for clean water SRF grants, as requested; $165 million, as requested, for U.S.-Mexico Border projects, Texas *colonias*, and Alaska Native Village projects; $450 million for drinking water SRF funding, contingent upon authorization; $674 million for state performance partnership consolidated management grants; and $129 million for seven special purpose grants.

In July, the Senate Appropriations Committee reported its version of H.R. 3666. The committee approved $2.815 billion for this account, consisting of $1.426 billion for clean water SRF grants; $550 million for drinking water SRF grants, contingent upon authorization; $165 million, as requested, for U.S.-Mexico border projects, Texas *colonias*, and Alaska Native Village projects; and $674 million for consolidated state grants. The committee rejected the provision of the House-passed bill providing $129 million for special purpose grants, including funds for Boston and New Orleans requested by the Administration, saying in report language that earmarking is provided at the expense of state revolving funds and does not represent an equitable distribution of grant funds (S.Rept. 104-318).

During debate on H.R. 3666 in September, the Senate adopted an amendment to reduce the FY1997 appropriation for clean water SRF grants by $725 million in order to fund the new drinking water SRF program. This action was intended to restore funds to the drinking water program which had been lost when Safe Drinking Water Act amendments were not enacted by August 1, 1996. Thus, the Senate-passed bill provided $701 million for clean water SRF grants and $1.275 billion for drinking water SRF grants for FY1997. Other amounts in the account were unchanged.

The conference report on H.R. 3666 (H.Rept. 104-812) was approved by the House and Senate on September 24, 1996. President Clinton signed the bill September 26 (P.L. 104-204). It reflected compromise of the House- and Senate-passed bills, providing the following amounts within the State and Tribal Assistance Grants account ($2.875 billion total):

- $625 million for clean water SRF grants,
- $1.275 billion for drinking water SRF grants,
- $165 million, as requested, for U.S.-Mexico border projects, Texas *colonias*, and Alaska Native Village projects,
- $136 million for 18 specific wastewater, water, and groundwater project grants (the 7 specified in House-passed H.R. 3666, plus 11 more; the bill provided funds for each of the needy cities projects requested by the Administration, but in lesser amounts), and
- $674 million for consolidated state grants, which could support implementation of a range of environmental programs.

The allocation of clean water and drinking water SRF grants was consistent with the Senate's action to restore funds to the drinking water program after enactment of the Safe Drinking Water Act amendments in early August.[19]

Subsequently, Congress passed a FY1997 Omnibus Consolidated Appropriations bill to cover agencies and departments for which full-year funding had not been enacted by October 1, 1996 (P.L. 104-208). It included additional funding for several EPA programs, as well as $35 million (on top of $40 million provided in P.L. 104-204) for the Boston Harbor cleanup project.

FY1998

President Clinton presented the Administration's budget request for FY1998 in February 1997. For water infrastructure and state and tribal assistance, the request totaled $2.793 billion, consisting of $1.075 billion for clean water SRF grants, $725 million for drinking water SRF grants, $715 million for consolidated state environmental grants, and $278 million for special project grants.

[19] P.L. 104-182, the Safe Drinking Water Act Amendments of 1996, enacted August 6, 1996, authorized the creation of the drinking water state revolving fund (DWSRF) program.

House and Senate committees began activities on FY1998 funding bills somewhat late in 1997, due to prolonged negotiations between Congress and the President over a five-year budget plan to achieve a balanced budget by 2002. After appropriators took up the FY1998 funding bills in June, the House passed EPA's appropriation in H.R. 2158 (H.Rept. 105-175) on July 15. In the State and Tribal Assistance Grants account, the House approved $3.019 billion, consisting of $1.25 billion for clean water SRF grants ($600 million more than FY1997 levels and $175 million more than requested by the President), $750 million for drinking water SRF grants ($425 million less than FY1997 levels, but $25 million more than the request), $750 million for state environmental assistance grants, and $269 million for special projects. The latter included funds for the special projects requested by the Administration but at reduced levels ($149 million total for these projects), plus $120 million in special project grants for 21 other communities.

The Senate passed a separate version of an FY1998 appropriations bill on July 22, 1997 (S. 1034, S.Rept. 105-53). It provided $3.047 billion for the STAG account, consisting of $1.35 billion for clean water SRF grants, $725 million for drinking water SRF grants, $725 million for state environmental assistance grants, and $247 million for special project grants. The Senate bill provided the amounts requested by the Administration for U.S.-Mexico border projects, Texas *colonias*, and Alaska Native Village projects (but no special funds for others requested by the President), plus $82 million for 18 special project grants for other communities identified in report language.

Conferees reached agreement on FY1998 funding in early October 1997 (H.R. 2158, H.Rept. 105-297). The final version passed the House on October 8 and passed the Senate on October 9. President Clinton signed the bill October 27 (P.L. 105-65). As enacted, it provided $3.213 billion for the STAG account, consisting of $1.35 billion for clean water SRF grants, $725 million for drinking water SRF grants, $745 million for consolidated state environmental assistance grants (which could address a range of environmental programs), and $393 million for 42 special purpose project and special community need grants for construction of wastewater,

water treatment and drinking water facilities, and groundwater protection infrastructure. It included the following amounts for grants requested by the Administration:

- $75 million for U.S.-Mexico border projects,
- $50 million for Texas *colonias*,
- $50 million for Boston Harbor wastewater needs,
- $10 million for New Orleans,
- $3 million for Bristol County, MA, and
- $15 million for Alaska Native Village projects.

The final bill also provided funds for all of the special purpose projects included in the separate House and Senate versions of the legislation, plus three projects not included in either earlier version.

Bill language was included in P.L. 105-65 to allow states to cross-collateralize clean water and drinking water SRF funds, that is, to use the combined assets of amounts appropriated to State Revolving Funds as common security for both SRFs, which conferees said is intended to ensure maximum opportunity for states to leverage these funds. Senate committee report language also said that the conference report on the 1996 Safe Drinking Water Act Amendments had stated that bond pooling and similar arrangements were not precluded under that legislation. The appropriations bill language was intended to ensure that EPA does not take an unduly narrow interpretation of this point which would restrict the states' use of SRF funds.[20]

On November 1, 1997, President Clinton used his authority under the Line Item Veto Act (P.L. 104-130) to cancel six items of discretionary budget authority provided in P.L. 105-65. The President's authority under this act took effect in the 105th Congress; thus, this was the first EPA appropriations bill affected by it. The cancelled items included funding for one of the special purpose grants in the bill, $500,000 for new water and sewer lines in an industrial park in McConnellsburg, PA. Reasons for the

[20] S.Rept. 105-53, accompanying S. 1034, 105th Cong., 1st sess., p. 71.

cancellation, according to the President, were that the project had not been requested by the Administration; it would primarily benefit a private entity and is outside the scope of EPA's usual mission; it is a low priority use of environmental funds; and it would provide funding outside the normal process of allocating funds according to state environmental priorities.[21]

However, in June 1998, the Supreme Court struck down the Line Item Veto Act as unconstitutional, and in July the Office of Management and Budget announced that funding would be released for 40-plus cancellations made in 1997 under that act (including those cancelled in P.L. 105-65) that Congress had not previously overturned.

FY1999

President Clinton's budget request for FY1999, presented to Congress in February 1998, requested $2.9 billion for the State and Tribal Assistance Grants account, representing 37% of the $7.9 billion total requested for EPA programs. The total included $1.075 billion for clean water SRF grants, $775 million for drinking water SRF grants, $115 million for water infrastructure projects along the U.S.-Mexico border projects and in Alaska Native Villages, $78 million for needy cities projects, and $875 million for consolidated state environmental grants (which could address a range of environmental programs).

Legislative action on the budget request occurred in mid-1998. Both houses of Congress increased amounts for water infrastructure financing, finding the Administration's request for clean water and drinking water SRF grants, as well as special project funding, not adequate. First, the Senate Appropriations Committee reported its version of an EPA spending bill in June 1998 (S. 2168, S.Rept. 105-216). This bill, passed by the Senate July 17, provided $3.2 billion for the STAG account, consisting of

[21] Office of Management and Budget. "Cancellation Pursuant to Line Item Veto Act," 62 *Federal Register* 59768, November 4, 1997. The President also cancelled funding for two other projects in the EPA portion of the bill, a water and wastewater training institute in Alabama and a solar aquatic wastewater treatment plant in Vermont. These projects were funded under a separate EPA account in the bill, the environmental programs and management account.

$1.4 billion for clean water SRF grants, $800 million for drinking water SRF grants, $105 million for U.S.-Mexico and Alaska Native Village projects, $100 million for 39 other special needs infrastructure grants, and $850 million for state performance partnership/categorical grants. As in FY1998, the committee included bill language allowing states to cross-collateralize their clean water and drinking water State Revolving Funds, making the language explicit for FY1999 and thereafter.

Second, the House passed its version of EPA's funding bill (H.R. 4194, H.Rept. 105-610) on July 29. This bill provided $3.2 billion for the STAG account, consisting of $1.25 billion for clean water SRF grants, $775 million for drinking water SRF grants, $70 million for U.S.-Mexico and Alaska Native Village projects, $253.5 million for 49 other special needs infrastructure grants (including nine projects also funded in the Senate bill), and $885 million for state environmental management grants (a 20% increase above FY1998 amounts for these state grants).

Conferees resolved differences between the two versions in October 1998 (H.R. 4194, H.Rept. 105-769). The conference agreement provided $3.4 billion for the STAG account, consisting of $1.35 billion for clean water SRF grants, $775 million for drinking water SRF grants, $80 million for U.S.-Mexico and Alaska Native and Rural Village projects, $301.8 million for 80 other special needs project grants, and $880 million for state and tribal environmental program grants (which could address a range of environmental programs). The House and Senate approved the agreement on October 7 and 8, respectively, and President Clinton signed the bill into law on October 21 (P.L. 105-276).

Additional funding was provided in the Omnibus Consolidated and Supplemental Appropriations Act, FY1999 (P.L. 105-277). This bill, which provided full-year funding for agencies and departments covered by seven separate appropriations measures, directed $20 million more in special needs grants for the Boston Harbor wastewater infrastructure project, on top of $30 million that was included in P.L. 105-276.

FY2000

For FY2000, beginning on October 1, 1999, the Administration requested $2.638 billion for water infrastructure assistance and state environmental grants. The total, $370 million less than the FY1999 appropriation for this account, consisted of $800 million for clean water SRF grants, $825 million for drinking water SRF grants, $128 million for Mexican border and special project grants, and $885 million for consolidated state environmental grants (which could address a range of environmental programs).

The request included one SRF policy issue. The Administration asked the appropriators to grant states the permission to set aside up to 20% of FY2000 clean water SRF monies in the form of grants for local communities to implement nonpoint source pollution and estuary management projects. Under the Clean Water Act, SRFs may only be used to provide loans. Some have argued that some types of water pollution projects which are eligible for SRF funding may not be suitable for loans, as they may not generate revenues which can be used to repay the loan to a state. This new authority, the Administration said, would allow states greater flexibility to address nonpoint pollution problems. Critics of the proposal said that making grants from an SRF would reduce the long-term integrity of a state's fund, since grants would not be repaid.

Some Members of Congress and stakeholder groups were particularly critical of the budget request for clean water SRF grants, $550 million (40%) less than the FY1999 level. Critics said the request was insufficient to meet the needs of states and localities for clean water infrastructure. In response, EPA acknowledged that several years prior the Clinton Administration had made a commitment to states that the clean water SRF would revolve at $2 billion annually in the year 2005. Because of loan repayments and other factors, EPA said, the overall fund will be revolve at $2 billion per year in the year 2002, even with the 20% grant setaside included in the FY2000 request. According to EPA, the $550 million decrease from 1999 would have only a limited impact on SRFs and would still allow the agency to meet its long-term capitalization goal of providing an average amount of $2 billion in annual assistance.

The House and Senate passed their respective versions of an EPA appropriations bill (H.R. 2684) in September 1999. The conference committee report resolving differences between the two versions (H.Rept. 106-379) was passed by the House on October 14 and the Senate on October 15 and was signed by the President on October 20 (P.L. 106-74). The final bill provided $7.6 billion overall for EPA programs, including $3.47 billion for the STAG account. Within that account, the bill included $1.35 billion for clean water SRF grants, $820 million for drinking water SRF grants, $885 million for categorical state grants (which generally support state and tribal implementation and could address a range of environmental programs), $80 million for U.S.- Mexico border and Alaska Rural and Native Village projects, and $331.6 million for 141 other special needs water and wastewater grants specified in report language. The final bill did not approve the Administration's request to allow states to use up to 20% of clean water SRF monies as grants for nonpoint pollution and estuary management projects.

Subsequent to enactment of the EPA funding bill, Congress passed the Consolidated Appropriations Act for FY2000 with funding for five other agencies (P.L. 106-113), which included provisions requiring a government-wide cut of 0.38% in discretionary appropriations. The bill gave the President some flexibility in applying this across-the-board reduction. Details of the reduction were announced at the time of the release of the FY2001 budget. EPA's distribution of the rescission resulted in a total reduction of $16.3 million for 139 of the special needs water and wastewater projects identified in P.L. 106-74. These projects were reduced 4.9% below enacted levels. The agency did not reduce funds for the two projects that had been included in the President's FY2000 budget request (Bristol County, MA, and New Orleans, LA) or for the United States-Mexico border and the Alaska Rural and Native Villages programs. EPA also reduced funds for the clean water SRF (enacted at $1.35 billion) by 0.3%, for a final funding level of $1.345 billion. The appropriation level was not reduced for the drinking water SRF or consolidated state grants.

FY2001

The President's budget for FY2001 requested a total of $2.9 billion for water infrastructure assistance and state environmental grants. For the second year in a row, President Clinton requested $800 million for the clean water SRF program, a $545 million reduction from the FY2000 level. The request included $825 million for the drinking water SRF program, $100 million for U.S.-Mexico border project grants, $15 million for Alaska Native Villages projects, two needy cities grants totaling $13 million (Bristol County, MA, and New Orleans, LA), plus $1.069 billion for consolidated state environmental grants (which could address a range of environmental programs).

The budget included a policy request similar to one in the FY2000 budget, which Congress rejected. The FY2001 budget sought flexibility for states to set aside up to 19% of clean water SRF monies in the form of grants for local communities to implement nonpoint source pollution and estuary management projects.

The House approved its version of EPA's funding bill (H.R. 4635, H.Rept. 106-674) on June 21, 2000. For the STAG account, H.R. 4635 provided $3.2 billion ($273 million more than requested, but $288 million below the FY2000 level). The total in the STAG account consisted of $1.2 billion for clean water SRF grants, $825 million for drinking water SRF grants, $1.068 billion (the budget request) for categorical state grants, and $85 million for U.S.-Mexico border and Alaska Rural and Native Villages projects. Beyond these, however, the House-passed bill included no funds for other special needs grants.

The Senate approved its version of the funding bill (S.Rept. 106-410) on October 12, 2000. For the STAG account, the Senate-passed bill provided $3.3 billion, consisting of $1.35 billion for clean water SRF grants, $820 million for drinking water SRF grants, $955 million for categorical state grants, $85 million for U.S.-Mexico border and Alaska Rural and Native Village projects, and $110 million for special needs water and wastewater grants.

In October, the House and Senate approved EPA's funding bill for FY2001 (H.Rept. 106-988), providing $1.35 billion for clean water SRF grants (the same level enacted for FY2000) and $825 million for drinking water SRF grants. The enacted bill included $110 million in grants for water infrastructure projects in Alaska Rural and Native Villages and U.S.-Mexico border projects and an additional $336 million for 237 other specified project grants throughout the country. The bill also provided $1,008 million for state categorical program grants ($60 million less in total than requested), which states could use to address a range of environmental programs. Total funding for the STAG account was $3.6 billion. Congress disapproved the Administration's policy request concerning use of clean water SRF monies for nonpoint source project grants. President Clinton signed the bill October 27, 2000 (P.L. 106-377).

Subsequently, in December, Congress provided $21 million more for five more special project water infrastructure grants (in addition to the $336 million in P.L. 106-377) as a provision of H.R. 4577, the FY2001 Consolidated Appropriations Act (P.L. 106-554). Also in that legislation, Congress enacted the Wet Weather Water Quality Act, authorizing a two-year, $1.5 billion grants program to reduce wet weather flows from municipal sewer systems. The provision was included in Section 112, Division B, of P.L. 106-554.

FY2002

In April 2001, the Bush Administration presented its budget request for FY2002. The Administration requested a total of $2.1 billion for clean water infrastructure funds, consisting of $823 million for drinking water SRF grants, $850 million for clean water SRF grants (compared with $1.35 billion appropriated for FY2001), and $450 million for the new program of municipal sewer overflow grants under legislation enacted in December, the Wet Weather Water Quality Act. However, that act provided that sewer overflow grants are only available in years when at least $1.35 billion in clean water SRF grants is appropriated. Subsequently, Administration officials said they would request that Congress modify the provision linking new grant funds to at least $1.35 billion in clean water SRF grants. The Bush budget requested no funds for special earmarked grants, except for $75 million to fund projects along the U.S.-Mexico border and $35 million for projects in Alaska Native Villages (both are the same amounts provided in FY2001). In response, some Members of Congress and outside groups criticized the budget request, saying that it did not provide enough support for water infrastructure programs. The President's budget also requested $1.06 billion for state categorical program grants, which generally support state and tribal administration of a range of environmental programs.

The House passed its version of FY2002 funding for EPA on July 30 (H.R. 2620, H.Rept. 107- 159). The House-passed bill provided a total of $2.4 billion for water infrastructure funds, consisting of $1.2 billion for clean water SRF grants, $850 million for drinking water SRF grants, $200 million for special project grants (individual projects were unspecified in the report accompanying H.R. 2620), $75 million for U.S.-Mexico border projects, and $30 million for Alaska Rural and Native Villages. The House bill provided no separate funds for the new wet weather overflow grant program, which the Administration had requested. Including $1.08 billion for state categorical program grants, total STAG account funding in the bill was $3.44 billion, about $150 million higher than the President's request.

The Senate passed its version of this appropriations bill on August 2 (S. 1216, S.Rept. 107-43). Like the House, the Senate rejected separate funding for wet weather overflow grants, and the Senate increased clean water SRF grant funding to the FY2001 level. The Senate-passed total for the STAG account was $3.49 billion, including $1.35 billion for clean water SRF grants, $850 million for drinking water SRF grants, $140 million for special needs infrastructure grants specified in accompanying report language, $75 million for U.S.-Mexico border projects, $30 million for Alaska Rural and Native Villages, and $1.03 billion for state categorical program grants.

Resolution of this and other appropriations bills in fall 2001 was complicated by congressional attention to general economic conditions and responses to the September 11 terrorist attacks on the World Trade Center and the Pentagon. Nevertheless, the House and Senate gave final approval to legislation providing EPA's FY2002 funding (H.R. 2620, H.Rept. 107-272) on November 8, and President Bush signed the bill on November 26 (P.L. 107-73). The final bill did not include separate funds for the new sewer overflow grant program requested by the Administration, which both the House and Senate had rejected, but it did include $1.35 billion for clean water SRF grants, $850 million for drinking water SRF grants, $344 million for 337 earmarked water infrastructure project grants specified in report language, and the requested $75 million for U.S.-Mexico border projects and $30 million for Alaska Rural and Native Villages. The bill included total STAG funding of $3.7 billion.

FY2003

President Bush presented the Administration's FY2003 budget request in February 2002, asking Congress to appropriate $2.185 billion for EPA's water infrastructure programs (compared with $2.659 billion appropriated for FY2002). The FY2003 request sought $1.212 billion for clean water SRF grants, $850 million for drinking water SRF grants, and $123 million for a limited number of special projects (especially in Alaska Native Villages and in communities on the U.S. Mexico border). The Administration proposed to eliminate funds for unrequested infrastructure project spending that Congress had earmarked in the FY2002 law, which totaled $344 million. Also, the Administration requested no funds for the municipal sewer overflow grants program enacted in 2000.

Some Members of Congress criticized the request level for clean water SRF capitalization grants, which was $138 million below the FY2002 enacted amount. In August 2002, the Senate Appropriations Committee approved an FY2003 funding bill for EPA that would provide $1.45 billion for clean water SRF grants, $100 million more than the FY2002 level (S. 2797, S.Rept. 107-222). In addition, the Senate committee bill included

$875 million for drinking water SRF grants, $140 million for special needs infrastructure grants specified in report language, $45 million for Alaska Rural and Native Village project grants, $75 million for U.S.-Mexico border projects, and $1.134 billion for state categorical program grants, which could address a range of environmental programs.

The House Appropriations Committee approved its version of an FY2003 funding bill with $1.3 billion for the clean water SRF program (H.R. 5605, H.Rept. 107-740) in October. This bill also included $850 million for drinking water SRF grants, $227.6 million for special needs infrastructure grants enumerated in report language, $35 million for Alaska Rural and Native Village project grants, $75 million for U.S.-Mexico border projects, and $1.173 billion for state categorical program grants, which could address a range of environmental programs. Neither appropriations committee included funds for the sewer overflow grant program authorized in 2000 (the Administration did not request FY2003 funds for these grants).

Due to complex budgetary disputes during the year, final action did not occur before the 107th Congress adjourned in November 2002, and it extended into 2003, more than five months after the start of the fiscal year. Congress and the President reached agreement on funding levels for EPA and other nondefense agencies in an omnibus appropriations act (P.L. 108-7; H.J.Res. 2, H.Rept. 108-10), which the President signed on February 20. The EPA portion of the enacted bill included $1.34 billion for clean water SRF grants, $844 million for drinking water SRF grants, and $413 million more for 489 special water infrastructure project grants to individual cities specified in conference report language, plus projects in Alaska Native Villages and communities on the U.S.-Mexico border. It also provided a total of $1.14 billion for categorical state grants, which generally support states and tribal implementation of a range of environmental programs.[22]

---

[22] H.J.Res. 2 included an across-the-board 0.65% reduction to accounts funded by the legislation, and to each program, project, and activity within an account. This reduction is reflected in amounts described here.

FY2004

On February 3, 2003, before completion of the FY2003 appropriations, President Bush submitted his budget request for FY2004. It requested a total of $1.798 billion for water infrastructure funds, consisting of $850 million for clean water SRF grants, $850 million for drinking water SRF grants, and $98 million for priority projects (especially in Alaska Native Villages and in communities on the U.S.-Mexico border). As in previous years, the Administration requested no funds for congressionally earmarked project grants for individual communities. Some Members of Congress and interest groups criticized the request for clean water SRF grants ($490 million below the FY2003 enacted level), but Administration officials responded by saying that the request reflected a commitment to fund this program at the $850 million level through FY2011. Funding at that level and over that long-term period, plus repayments of previous SRF loans made by states, would be expected to increase the revolving levels of the overall program from $2.0 billion to $2.8 billion per year, the Administration said. The President's budget also requested $1.2 billion for categorical state grants, which could address a range of environmental programs.

On July 25, the House approved H.R. 2861 (H.Rept. 108-235), providing FY2004 appropriations for EPA. As passed, the bill included $1.2 billion for clean water SRF grants, $850 million for drinking water SRF grants, $203 million for earmarked water infrastructure project grants, and $75 million in grants for high-priority projects in Alaska Native Villages and along the U.S. Mexico border. Senate action on its version of a funding bill for EPA (S.Rept. 108-143) occurred on November 18. The Senate-passed bill provided $1.35 billion for clean water SRF grants, $850 million for drinking water SRF grants, $130 million for targeted infrastructure project grants, plus $95 million in grants for projects in Alaska Native Villages and along the U.S.-Mexico border.

As with the previous year's appropriations, Congress did not enact legislation providing FY2004 funds for EPA before the beginning of the new fiscal year; thus EPA programs were covered by a series of continuing resolutions (CRs). The last of these CRs (P.L. 108-135) extended FY2003 funding levels through January 31, 2004. On December 8, 2003, the House passed legislation providing full-year funding for EPA and other agencies that lacked enacted appropriations (H.R. 2673). The conference report on this bill (H.Rept. 108-401) provided $1.34 billion for clean water SRF grants, $845 million for drinking water SRF grants, and $425 million in grants for 520 earmarked grants in listed communities, Alaska Native Villages, and U.S.-Mexico border projects.[23] The Senate approved the conference report on January 22, 2004, and President Bush signed the legislation January 23 (P.L. 108-199).

FY2005

The FY2005 EPA appropriation for water infrastructure funds was the lowest total for these programs since FY1997 (the first year in which Congress provided both clean water and drinking water SRF capitalization grants, as well as earmarked project grants). The decline was due primarily to a reduction in funding for the clean water SRF program from an average of $1.35 billion since FY1998 to $1.09 billion.

President Bush's FY2005 budget, presented February 2, 2004, requested a total of $3.0 billion for water infrastructure assistance and state environmental program grants. It included $850 million for clean water SRF grants, $850 million for drinking water SRF grants, $94 million for priority projects (primarily in Alaska Native Villages and along the U.S.-Mexico border), and $1.25 billion for categorical grants, which could address a range of environmental programs. As in recent budgets, the Administration requested no funds for congressionally earmarked project grants. Anticipating that critics likely would focus on the clean water SRF request ($492 million below the FY2004 level), in its budget documents the Administration said that the request included funding for the clean

[23] H.R. 2673 included an across-the-board 0.59% reduction to accounts funded by the legislation, and to each program, project, and activity within an account. This reduction is reflected in amounts described here.

water SRF at $850 million annually through 2011, which, together with loan repayments, state matches, and other funding sources, would result in a long-term average revolving level of $3.4 billion. Likewise, the budget anticipated funding the drinking water SRF program at the same $850 million annually through 2011, resulting in a long-term average revolving level of $1.2 billion.

House and Senate Appropriations committees began review of the EPA budget request in March. On September 9, 2004, the House Appropriations Committee reported FY2005 funding for EPA in a bill that included the Administration's requested level of $850 million for clean water SRF grants, $850 million for drinking water SRF grants, and earmarked grants for priority water infrastructure projects totaling $393.4 million (H.R. 5041, H.Rept. 108-674). On September 21, the Senate Appropriations Committee reported its version of this bill (S. 2825, S.Rept. 108-353), which included $1.35 billion for clean water SRF grants, $850 million for drinking water SRF grants, and $217 million for earmarked project grants.

Final action on the FY2005 appropriation did not occur before the start of the fiscal year. On November 20, the House and Senate passed H.R. 4818 (H.Rept. 108-792), the Consolidated Appropriations Act, 2005, an omnibus appropriations bill comprising nine appropriations measures, including funding for EPA. The bill provided total funding for EPA of $8.1 billion,[24] a decrease from the $8.4 billion approved in FY2004, but $340 million more than was requested by the President in February. One of the most controversial items in the final bill was a $251 million decrease for clean water SRF grants from the FY2004 level, although the $1.09 billion total was $241 million more than in the President's budget. The final measure also included $843 million for drinking water SRF capitalization grants; $401.7 million for 669 earmarked grants in listed communities, Alaska Native Villages, and U.S.-Mexico border projects; and $1.14 billion for categorical state grants, which generally support state and tribal

[24] H.R. 4818 included an across-the-board 0.80% reduction to accounts funded by the legislation, and to each program, project, and activity within an account. This reduction is reflected in amounts described here.

administration of a range of environmental programs. The $2.34 billion total for water infrastructure programs and projects was $542 million more than was requested by the President, but $276 million less than Congress appropriated for FY2004. President Bush signed the legislation December 8, 2004 (P.L. 108-447).

FY2006

The FY2006 appropriation for water infrastructure funds marked the second consecutive year in which Congress appropriated less funding for these programs, providing lower levels both for clean water SRF capitalization grants and for earmarked project grants than in FY2005.

President Bush presented the FY2006 budget request in February 2005. Overall for EPA, it sought 5.6% less than Congress had appropriated for FY2005. The Administration's deepest cuts affecting EPA were proposed for the STAG account. The budget requested $730 million for clean water SRF grants (33% below FY2005 appropriated funding and 45.6% below the FY2004 level), $850 million for drinking water SRF grants (a slight increase from the FY2005 level), $69 million for priority projects (primarily in Alaska Native Villages and along the U.S.-Mexico border), and $1.2 billion for state categorical grants, which could address a range of environmental programs. As in previous years, the Administration requested no funds for congressionally earmarked water infrastructure projects. Advocates for the SRF programs (especially state and local government officials) contended that cuts to the clean water program would impair their ability to carry out needed municipal wastewater treatment plant improvement projects. Administration officials responded that the proposed SRF reductions for FY2006 were because Congress had boosted funds above the FY2005 request level. These officials said that the Administration planned to invest $6.8 billion in the clean water SRF program between FY2004 and FY2011, after which federal funding was expected to end, and the state SRFs were expected to have an annual revolving level of $3.4 billion. If Congress appropriated more than requested in any given year (as occurred in FY2005), they said, that target would be met sooner, leading to reduced requests for the SRF in subsequent years until a planned phase-out in FY2011.

On May 19, 2005, the House passed H.R. 2361, providing FY2006 funding for EPA. As passed, it provided $850 million for clean water SRF grants ($120 million more than the President's request), $850 million for drinking water SRF grants, and $269 million for earmarked water infrastructure grants. During debate, the House rejected two amendments to increase clean water SRF funding. On June 29, the Senate passed its version of H.R. 2361, providing $1.1 billion for clean water SRF grants, $850 million for drinking water SRF grants, and $290 million for earmarked project grants. The House bill required that $100 million of the SRF funding come from balances from expired contracts, grants, and interagency agreements from various EPA appropriation accounts. The Senate bill, in contrast, called for a $58 million rescission of unobligated amounts associated with grants, contracts, and interagency agreements in various accounts, but did not specify that such monies go to SRF funding.

Conferees resolved differences between the bills (H.Rept. 109-188), and the House and Senate approved the measure in July; the President signed it into law on August 2 (P.L. 109-54). As enacted, the bill provided $900 million for clean water SRF grants; $850 million for drinking water SRF grants; $285 million for 259 earmarked grants in listed communities, Alaska Native Villages, and along the U.S.-Mexico border; and $1.13 billion for categorical state grants, which could address a range of environmental programs. The final bill required an $80 million rescission from expired grants, contracts, and interagency agreements in various EPA accounts (not just the STAG account) not obligated by September 1, 2006. It did not direct the rescinded funds to be applied to the clean water SRF, as proposed by the House. The $2.03 billion total in the bill for EPA water infrastructure programs and projects was $386 million more than was requested by the President, but $301 million less than Congress appropriated for FY2005.

Further, the funding amounts specified in P.L. 109-54 were reduced slightly. First, a provision of P.L. 109-54, Section 439, mandated an across-the-board rescission of 0.476% for any discretionary appropriation in that bill. Second, in December 2005 Congress enacted P.L. 109- 148, the FY2006 Department of Defense Appropriations Act, and Section 3801 of that bill mandated a 1% across-the-board rescission for discretionary accounts in any FY2006 appropriation act (except for discretionary authority of the Department of Veterans Affairs). As a result of these two rescissions, the final levels for the STAG account were $887 million for clean water SRF grants; $838 million for drinking water SRF grants; $281 million for 259 earmarked grants in listed communities, Alaska Native Villages, and along the U.S.-Mexico border; and $1.11 billion for categorical state grants, which could address a range of environmental programs. FY2006 EPA water infrastructure programs and projects thus total $2.0 billion.

On October 28, President Bush requested that Congress rescind $2.3 billion from 55 "lower-priority federal programs and excess funds," including $166 million from clean water SRF monies. In the end, Congress did not endorse the specific request to reduce clean water SRF appropriations. The two rescissions resulting from P.L. 109-54 and P.L. 109-148 totaled a $13.2 million reduction from the $900 million specified in the EPA appropriations act.

FY2007

President Bush presented the Administration's FY2007 budget request in February 2006, asking Congress to appropriate $1.570 billion for EPA's water infrastructure programs. The FY2007 request sought $687.6 million for clean water SRF grants, $841.5 million for drinking water SRF grants, and $40.6 million for special projects in Alaska Native Villages, Puerto Rico, and along the U.S.-Mexico border. When the 109th Congress adjourned in December 2006, it had not completed action on appropriations legislation to fund EPA (or on nine other appropriations bills covering the majority of domestic discretionary agencies and departments) for the fiscal year that began October 1, 2006, thus carrying over this legislative activity into the 110th Congress. In December 2006, Congress enacted a continuing resolution, P.L. 109-383 (the third such continuing resolution since the start of the fiscal year on October 1), providing funds for EPA and the other affected agencies and departments until February 15, 2007.

The President's FY2007 budget request for clean water SRF capitalization grants was 22% less than the FY2006 appropriation for these grants and 37% below the FY2005 funding level. The request for drinking water SRF grants was essentially the same as in recent years ($4 million more than FY2006, $1.7 million less than FY2005). As in recent budgets, the Administration proposed no funding for congressionally designated water infrastructure grants, but, as noted above, it did seek a total of $40.6 million for Administration priority projects. Advocates of the clean water SRF program (especially state and local government officials) again contended, as they have for several recent years, that the cuts would impair their ability to carry out needed municipal wastewater treatment plant improvement projects. Administration officials responded that cuts for the clean water SRF in FY2007 were necessary because Congress boosted funds above the requested level in FY2005 and FY2006.

On May 18, 2006, the House passed H.R. 5386 (H.Rept. 109-465), providing the requested level of $687.6 million for clean water SRF grants and $841.5 million for drinking water SRF grants. The Senate Appropriations Committee approved the same funding levels for these grant programs when it reported H.R. 5386 on June 29 (S.Rept. 109-275), but the Senate did not act on this measure before the 109th Congress adjourned in December. Before adjournment, Congress enacted a continuing resolution (CR), P.L. 109-383 (the third such CR since the start of the fiscal year on October 1), providing funds for EPA and the other affected agencies and departments until February 15, 2007. Funding levels provided under this CR followed a "lowest level" concept for individual programs; that is, programs were funded at the lowest level under either House-passed FY2007 appropriations, Senate-passed appropriations, or the

FY2006 funding. For clean water SRF grants, the resulting appropriation through mid-February was $687.6 million, as in House-passed H.R. 5386. For drinking water SRF grants, the appropriation level through mid-February was $837.5 million, the FY2006-enacted level. The CR included funds for congressionally earmarked water infrastructure project grants totaling $200 million, as in House-passed H.R. 5386.

Returning to these issues in 2007, in mid-February, Congress passed H.J.Res. 20, a continuing appropriations resolution that provides funding for EPA and the other affected agencies through the end of FY2007. As passed, this full-year resolution held most programs and activities at their FY2006 appropriated levels. However, clean water SRF capitalization grants were one of the few programs that received a funding increase under the resolution: these grants received $1.08 billion ($197 million more than in FY2006, and $396 million more than the President requested for FY2007). The resolution further prohibited project grants for congressional earmarks, but not for special project grants requested in the President's budget. The action to ban earmarks in FY2007 occurred when leaders in the 110th Congress sought to finish up appropriations actions that were unresolved at the end of the 109th Congress, and at the same time the newly elected Congress moved to adopt rules and procedures to reform the congressional earmarking process for the future. (Water infrastructure project earmarks totaled $281 million in EPA's FY2006 appropriation.) President Bush signed H.J.Res. 20 on February 15, 2007 (P.L. 110-5).

The final FY2007 amounts provided in P.L. 110-5 were

- $1.084 billion for clean water SRF capitalization grants,
- $837.5 million for drinking water SRF capitalization grants,
- $83.75 million for Alaska Native Village and U.S.-Mexico border project grants requested by the Administration, and
- $1.11 billion for categorical state grants, which could be used to administer a range of environmental programs.

FY2008

President Obama presented his FY2008 budget request to Congress on February 5, 2007, before finalization of the FY2007 appropriations. The budget sought $687.6 million for clean water SRF grants, the same amount requested for FY2007; $842.2 million for drinking water SRF grants; $25.5 million for special project grants for Alaska Native Villages and the U.S.-Mexico border region; and $1.065 billion for categorical state grants, which could address a range of environmental programs.

In June 2007, the House passed H.R. 2643, providing FY2008 appropriations for EPA. This bill included $1.125 billion for clean water SRF grants, $842.2 million for drinking water SRF grants, plus $175.5 million for 143 congressionally designated water infrastructure project grants. The Senate Appropriations Committee approved companion legislation (S. 1696) that similarly included higher funding levels for several water quality programs. The Senate committee's bill provided less funding for clean water SRF grants than the House bill ($887 million), the same amount for drinking water SRF grants, and slightly more for congressionally designated water infrastructure project grants ($180 million). The Senate did not take up S. 1696.

By October 1, the start of FY2008, Congress had not enacted any appropriations bills for FY2008, and Congress enacted several short-term continuing appropriations resolutions to temporarily fund EPA and other government agencies until final agreement, which occurred in December 2007. Full-year funding for EPA's water infrastructure programs was included in the Consolidated Appropriations Act for FY2008 (Division F, Title II), signed by the President December 26, 2007 (P.L. 110-161).

The final FY2008 amounts[25] provided in this legislation were

- $689.1 million for clean water SRF capitalization grants ($1.5 million more than requested by the Administration),

[25] Finals amounts shown here reflect a 1.56% across-the-board reduction of appropriated amounts for accounts included in Division F of the legislation.

- $829.0 million for drinking water SRF capitalization grants ($13.2 million less than requested),
- $177.2 million for 282 earmarked grants in listed communities, Alaska Native Villages, and U.S.-Mexico border projects ($151.7 million more than requested), and
- $1.078 billion for categorical state grants ($13.3 million more than requested), which could address a range of environmental programs.

FY2009

President Obama presented his FY2009 budget request to Congress on February 6, 2008. The budget sought $555 million for clean water SRF grants, $134 million less than Congress appropriated for FY2008; $842.2 million for drinking water SRF grants, $13 million more than was appropriated for FY2008; $25.5 million for special project grants for Alaska Native Villages and the U.S.-Mexico border region, $18.8 million less than was appropriated for FY2008; and $1.057 billion for categorical state grants, which could address a range of environmental programs. As in past years, the budget requested no funds for other earmarked grants.

In June 2008, a House Appropriations subcommittee approved a bill with FY2009 funding for EPA, but no further action occurred before the start of the fiscal year. At the end of September 2008, Congress and the President agreed to legislation providing partial-year funding for EPA and most other agencies and departments. This bill, the Consolidated Security, Disaster Assistance, and Continuing Resolution Act, 2009 (P.L. 110-329), provided funding through March 6, 2009, at FY2008 funding levels. A second short-term continuing resolution was enacted on March 6 (P.L. 111-6), while Congress was finishing consideration of a full-year omnibus FY2009 appropriations bill that the President signed on March 11 (P.L. 111-8). The omnibus bill provided $689 million in regular appropriations for clean water SRF grants, $829 million for drinking water SRF grants—both at the same levels as were appropriated in FY2008—and $1.094

billion for categorical state grants, which support administration of a range of environmental programs. The omnibus appropriations act also includes $183.5 million for earmarked water infrastructure grants.

FY2009 Supplemental Appropriations, the American Recovery and Reinvestment Act

In February 2009, Congress responded to the nation's economic crisis by enacting the American Recovery and Reinvestment Act (ARRA, P.L. 111-5), legislation providing FY2009 supplemental appropriations to a number of government programs. Part of the philosophy underlying the legislation was the concept of using federal investments to make accelerated investments in the nation's public infrastructure in order to create jobs while also meeting infrastructure needs. To that end, the legislation included $4.0 billion for clean water SRF capitalization grants (for total FY2009 funds of $4.689 billion) and $2.0 billion for drinking water SRF capitalization grants (for total FY2009 funds of $2.829 billion). The supplemental SRF funds were available for obligation through FY2010, but under the legislation, states were to give preference when awarding funds to activities that can start and finish quickly, with a goal that at least 50% of the funds go to activities that can be initiated within 120 days of enactment. States were to give priority to wastewater projects that could proceed to construction within 12 months of enactment, and funds for projects that were not under contract or under construction by February 12, 2010, would be reallocated by EPA to other states. Further, the legislation required states to reserve at least 20% of the SRF capitalization grant funds for a Green Project Reserve, that is, projects intended to achieve improved energy or water efficiency. It also specified that all assistance agreements made in whole or in part with funds appropriated under the ARRA must comply with prevailing wage requirements of the Davis-Bacon Act.[26]

---

[26] For additional information, see CRS Report R40216, *Water Infrastructure Funding in the American Recovery and Reinvestment Act of 2009*, by Claudia Copeland, Megan Stubbs, and Charles V. Stern.

FY2010

President Obama presented his Administration's FY2010 budget request on May 7, 2009. For EPA as a whole, the budget sought $10.5 billion, a 38% increase above levels enacted in EPA's regular FY2009 appropriations (P.L. 111-8). The bulk of the increase in the President's budget was for water infrastructure assistance, which would receive 157% above FY2009 levels (excluding ARRA supplemental funds). The request included

- $2.4 billion for clean water SRF capitalization grants;
- $1.5 billion for drinking water SRF capitalization grants;
- $20 million for Alaska Native Village and U.S.-Mexico border projects; and
- $1.111 billion for state categorical grants (1.5% above FY2009 levels), which generally support state administration of environmental programs.

Congress provided FY2010 appropriations for EPA in P.L. 111-88, passed by the House and Senate in October 2009 and signed into law on October 30. In this measure, Congress provided the following:

- $2.1 billion for clean water SRF capitalization grants;
- $1.387 billion for drinking water SRF capitalization grants;
- $186.7 million for 319 congressionally earmarked special project grants, including assistance for Alaska Native Villages and U.S.-Mexico border projects; and
- $1.116 billion for state categorical environmental grants, which could address a range of environmental programs.

The FY2010 appropriations act included some restrictions that Congress also had specified in the American Recovery and Reinvestment Act, discussed above, namely a requirement that 20% of SRF capitalization grant assistance be used for "green" infrastructure and also that Davis-

Bacon Act prevailing wage rules shall apply to construction of wastewater or drinking water projects carried out in whole or in part with assistance from the SRF.

FY2011

President Obama presented the FY2011 budget request in February 2010. For EPA as a whole, the budget sought $10.02 billion in discretionary budget authority, a 3% decrease from levels enacted for EPA in FY2010. The largest component of the reduced request, compared with FY2010, was $200 million less for grants to capitalize clean water and drinking water SRF programs. In explaining the request, EPA budget documents noted that even with a slight reduction, the budget "continues robust funding for the SRFs." As in past years, the President requested no funds for congressionally designated water infrastructure projects.[27] The request included

- $2.0 billion for clean water SRF capitalization grants;
- $1.287 billion for drinking water SRF capitalization grants;
- $20 million for Alaska Native Village and U.S.-Mexico border projects; and
- $1.277 billion for state categorical grant programs (14% higher than the FY2010 enacted amount), which could address a range of environmental programs.

Congress took only limited action on FY2011 funding for EPA before the start of the new fiscal year on October 1, 2010: a House Appropriations subcommittee approved a bill in July, but no further action followed. At the end of September, the House and Senate passed a continuing resolution to extend FY2010 funding levels for EPA and other federal agencies and departments until December 3, 2010, because no FY2011 appropriations

[27] For additional information, see CRS Report R41149, *Environmental Protection Agency (EPA): Appropriations for FY2011*, by Robert Esworthy et al.

bills had been enacted by October 1. President Obama signed the continuing resolution (CR) on September 30 (P.L. 111-242). This bill was followed by six more short-term CRs before Congress came to final resolution of FY2011 spending on April 14, 2011, enacting a bill to provide funding for EPA and all other federal agencies and departments through September 30 (P.L. 112-10). The final bill reduced overall funding for EPA 15% below the FY2010 level.

The enacted bill included[28]

- $1.522 billion for clean water SRF capitalization grants;
- $963.1 million for drinking water SRF capitalization grants;
- $19.96 million for Alaska Native Village and U.S. Mexico-border projects; and
- $1.254 billion for state categorical grant programs, which generally support implementation of a range of environmental programs.

FY2012

Policymakers began to consider the budget for FY2012 before finalizing the funding levels for FY2011. The President submitted the Administration's FY2012 budget request on February 14, 2011. It sought $9 billion total for EPA, a decrease of $1.3 billion from the FY2010 enacted level, but 3% higher than the FY2011 enacted level. The President's request included $1.55 billion for clean water SRF capitalization grants, $990 million for drinking water SRF capitalization grants, $20 million for Alaska Native Village and U.S.-Mexico border assistance, and $1.2 billion for state categorical grants, which could address a range of environmental programs.

For several days in July 2011, the House debated H.R. 2584, providing FY2012 appropriations for EPA, but did not take final action on the bill

---

[28] These amounts reflect a mandated 0.2% across-the-board rescission that was included in the final bill.

before the August recess. As reported, the bill provided $7.3 billion for EPA, 17% less than FY2011 funds and 19% less than the President's FY2012 request. It reduced funds for the clean water SRF capitalization grants to $689 million and $829 million for drinking water SRF capitalization grants (the same levels provided in FY2008), while including no funds for congressionally designated special projects (i.e., earmarks). The reported bill also provided $1.002 billion for state categorical grants, which could address a range of environmental programs. There was no action on this bill in the Senate.

Final congressional action on FY2012 appropriations for EPA and most other federal agencies and departments did not occur until the end of December 2011, enacted in an omnibus appropriations act, P.L. 112-74. The enacted bill included[29]

- $1.466 billion for clean water SRF capitalization grants (3.7% below FY2011);
- $917.9 million for drinking water SRF capitalization grants (4.7% below FY2011);
- $14.976 million for Alaska Native Village and U.S.-Mexico border projects; and
- $1.089 billion for state categorical grants, which could address a range of environmental programs.

FY2013

President Obama presented the Administration's FY2013 budget request in February 2012. It sought $8.34 billion overall for EPA, or 4.7% below the level enacted for FY2012. The request included $1.175 billion for clean water SRF capitalization grants, $850 million for drinking water SRF capitalization grants, $20 million for Alaska Native Village and U.S.-Mexico border assistance, and $1.2 billion for state categorical grants,

[29] These amounts reflect a mandated 0.16% across-the-board rescission for programs in EPA's portion of the legislation that was included in the final bill.

which could address a range of environmental programs. The total amount requested for SRF capitalization grants is 15% below the FY2012 enacted level, reflecting a 20% reduction for the clean water program and a 7.4% reduction for the drinking water program.

The House Appropriations Committee approved legislation providing FY2013 funds for EPA in July 2012 (H.R. 6091). As reported, the bill provided $689 million for clean water SRF capitalization grants (the same level provided in FY2008), $829 million for drinking water SRF capitalization grants, $994 million for state categorical grants, and no funds for Alaska Native Village or U.S.-Mexico border projects.

The House did not take up H.R. 6091, nor did the Senate act on an EPA appropriations bill (although the Senate Appropriations Committee released a draft bill in September 2012).[30] Prior to the start of FY2013 on October 1, 2012, Congress passed and the President signed a continuing resolution bill providing funding for government agencies and departments through March 27, 2013 (P.L. 112-175). This measure funded the government generally at FY2012 levels plus a 0.6% increase.

Final action on FY2013 appropriations occurred in the Further Continuing Appropriations Act, 2013 (P.L. 113-6). Funding enacted in this bill included $1.452 billion for clean water SRF capitalization grants; $908.7 million for drinking water SRF capitalization grants; $15 million for Alaska Native Village and U.S.-Mexico border assistance; and $1.1 billion for state categorical grants, which generally support state and tribal implementation of a range of environmental programs. However, these amounts were reduced under the March 1, 2013, sequester order of the President, which reduced affected accounts by 5.0%, and by an across-the-board rescission of 0.2% necessary to avoid exceeding the FY2013 discretionary spending limits in law. After these reductions, available FY2013 funding was approximately $1.38 billion for the clean water SRF capitalization grants, $860 million for drinking water SRF capitalization

---

[30] See http://www.appropriations.senate.gov/news.cfm?method=news.view&id=fc23708b-fb33-4569-99b4-6cf0d0254457.

grants, $14 million for Alaska Native Village and U.S.-Mexico border assistance, and $1.0 billion for state categorical grants.[31]

FY2014

President Obama presented the Administration's FY2014 budget in April 2013. It sought $8.15 billion overall for EPA, including $1.095 billion for clean water SRF capitalization grants, $817 million for drinking water SRF capitalization grants, $15 million for Alaska Native Village and U.S.-Mexico border projects, and $1.136 billion for state categorical grants. The total amount requested for SRF capitalization grants was 19% below the FY2013 enacted level.

In mid-2013, the House Appropriations Subcommittee on Interior, Environment, and Related Agencies drafted a bill (unnumbered) that would have reduced overall funding for EPA by 34% from the FY2013 enacted level, including an 83% reduction for clean water SRF capitalization grants (the bill would have provided $250 million) and a 65% reduction for drinking water SRF capitalization grants ($350 million was included in the bill). According to subcommittee documents, the reduction was appropriate because, despite recent federal support, little progress has been made to reduce the known water infrastructure gap.[32] The full committee did not complete markup of this bill.

The Senate Appropriations Subcommittee on Interior, Environment, and Related Agencies drafted an alternative bill that would have maintained funding for the clean water SRF program at $1.45 billion and funding for the drinking water SRF program at $907 million.[33] There was no further action on this bill.

Congress did not reach final agreement on FY2014 appropriations before the start of the fiscal year on October 1, but did agree to a short-

---

[31] Personal communication, U.S. Environmental Protection Agency.

[32] See https://appropriations.house.gov/uploadedfiles/hrpt-113-hr-fy2014-interior.pdf p. 62.

[33] See http://www.appropriations.senate.gov/news.cfm?method=news.view&id=b3e22f9d-a060-45eb-90ef%C2%AC1225244125a7.

term continuing appropriations measure (P.L. 113-46), which provided funding through January 15, 2014. Final action on appropriations for EPA and all other federal agencies and departments occurred as part of the Consolidated Appropriations Act, 2014 (H.R. 3547, P.L. 113-76), signed by the President on January 17, 2014. This bill provides $1.45 billion for clean water SRF capitalization grants (5% more than FY2013 funds and 32% higher than the President's FY2014 budget request) and $907 million for drinking water SRF capitalization grants (5% more than FY2013 funds and 11% higher than the President's FY2014 budget request). The bill also provides $15 million for Alaska Native Village and U.S.-Mexico border assistance, and $1.0 billion for state categorical grants, which generally support state and tribal implementation of a range of environmental programs.

FY2015

President Obama presented the Administration's FY2015 budget on March 4, 2014. It sought $7.89 billion overall for EPA, including $1.018 billion for clean water SRF capitalization grants, $757 million for drinking water SRF capitalization grants, $15 million for Alaska Native Village and U.S.-Mexico border projects, and $1.13 billion for state categorical grants. The total amount requested for SRF capitalization grants was 25% below the FY2014 enacted level.

Final full-year appropriations were enacted as part of the Consolidated and Further Continuing Appropriations Act, 2015, enacted in December 2014 (P.L. 113-235). The legislation provided the same water infrastructure funding levels as in FY2014: $1.45 billion for clean water SRF capitalization grants and $907 million for drinking water SRF capitalization grants. As with the FY2014 appropriations, the bill provided $15 million for Alaska Native Village and U.S.-Mexico border assistance and $1.0 billion for state categorical grants, which could address a range of environmental programs.

FY2016

The Administration's FY2016 budget requested $8.6 billion overall for EPA. The request included $1.116 billion for clean water SRF capitalization grants, $1.186 billion for drinking water SRF capitalization grants (31% higher than the FY2016 appropriation), $15 million for Alaska Native Village and U.S.-Mexico border projects, and $1.162 billion for state categorical grants, which generally support state and tribal implementation of a range of environmental programs.

Although the House and Senate Appropriations Committees reported bills to provide FY2016 appropriations for EPA, final appropriations action for EPA and other agencies occurred as part of the Consolidated Appropriations Act, 2016, signed by the President December 18, 2015 (P.L. 114- 113). The bill provided $1.394 billion for clean water SRF capitalization grants ($55 million less than FY2015, but $278 million above the President's request), $863 million for drinking water SRF capitalization grants ($44 million below the FY2015 level, and $323 million less than the President's request), and $30 million for Alaska Native Village and U.S.-Mexico border water infrastructure projects. It also provided $1.06 billion for state categorical grants.

FY2017

President Obama presented the Administration's FY2017 budget in February 2016, requesting $8.3 billion in total for EPA ($127 million above the FY2016 enacted budget). The request for EPA included $979.5 million for clean water SRF capitalization grants ($424 million less than the FY2016 enacted level), $1.02 billion for drinking water SRF capitalization grants ($157 million above the FY2016 amount), $22 million for Alaska Native Village and U.S.-Mexico border projects, and $1.158 billion for state categorical grants, which generally support state and tribal implementation of environmental programs.

During congressional hearings on the EPA request, many Members criticized the requested 30% decrease in funds for clean water SRF capitalization grants. This criticism was reflected to some degree in appropriations bills the Appropriations Committees subsequently approved that include EPA funding. In July 2016, the House passed H.R. 5538, FY2017 Interior and Environment Appropriations Act; it included $1.0 billion for clean water SRF grants, $1.07 billion for drinking water SRF grants, and $1.06 billion for state categorical grants. The Senate Appropriations Committee reported a companion bill, S. 3068, in June. It included $1.35 billion for clean water SRF grants, $1.02 billion for drinking water SRF grants, and $1.09 billion for state categorical grants. The Senate did not take up this bill.

Congress did not reach final agreement on an EPA funding bill before the start of FY2017. However, on September 28, the House and Senate passed a 10-week continuing resolution that extended FY2016 funding levels, minus a 0.496% across-the-board reduction, through December 9, 2016 (P.L. 114-223). A second continuing resolution, passed in December 2014, extended FY2016 funding levels, minus a 0.1901% across-the-board reduction, from December 10, 2016, through April 28, 2017 (P.L. 114-254).

The Obama Administration's FY2017 budget submission also included a $15 million request to allow EPA to begin making water infrastructure project loans under a program that Congress enacted in 2014, the Water Infrastructure Financing and Investment Act, or WIFIA.[34] P.L. 114 254 included the first appropriation, $20 million, for EPA to do so. The FY2017 final appropriations act (discussed below) provided an additional $8 million for EPA's WIFIA program (and $2 million for EPA to administer the program). The Continuing Appropriations Act, 2018, and Supplemental Appropriations for Disaster Relief Requirements Act, 2017 (P.L. 115-56, §133), enacted on September 8, 2017, provided an additional $3 million to EPA for administrative expenses.

---

[34] For more information, see CRS Report R43315, *Water Infrastructure Financing: The Water Infrastructure Finance and Innovation Act (WIFIA) Program*, by Jonathan L. Ramseur and Mary Tiemann.

Final full-year appropriations were enacted as part of the Consolidated and Further Continuing Appropriations Act, 2017, signed by President Trump on May 5, 2017 (P.L. 115-31). The act provided the same level of funding for water infrastructure as FY2016: $1.394 billion for clean water SRF capitalization grants ($414 million above President Obama's request), $863 million for drinking water SRF capitalization grants ($158 million less than President Obama's request), and $30 million for Alaska Native Village and U.S.-Mexico border water infrastructure projects. It also provided $1.07 billion for state categorical grants, which support a range of environmental programs.

The Continuing and Security Assistance Appropriations Act, 2017 (P.L. 114-254), included an additional $100 million in DWSRF funding to assist Flint, MI, as authorized in the WIIN Act (P.L. 114-322).

FY2018

The Trump Administration's FY2018 budget request proposed $8.6 billion overall for EPA. The request included $1.394 billion for clean water SRF capitalization grants and $863 million for drinking water SRF capitalization grants (the same amounts as the FY2017 appropriation). The request proposed $597 million for state categorical grants, a 44% reduction compared to FY2017 levels. Much of this reduction came from the elimination of funding for nonpoint source grants (CWA Section 319)[35] and reduction of grant funding for water pollution control (CWA Section 106).[36] In addition, the President's budget request proposed to eliminate funding for Alaska Native Village and U.S.-Mexico border projects.

Similar to the previous fiscal year, Congress did not reach final agreement on an EPA funding bill before the start of FY2018. EPA and other federal departments and agencies operated under multiple continuing resolutions generally at FY2017 enacted levels (minus across-the-board rescissions). Final full-year appropriations were enacted as part of the

[35] 33 U.S.C. §1329.
[36] 33 U.S.C. §1256.

Consolidated Appropriations Act, 2018, signed by President Trump on March 23, 2018 (P.L. 115-141). EPA's STAG account (Division G, Title II) included $1.394 billion for the clean water SRF and $863 million for the drinking water SRF program (the same amounts appropriated for FY2017, less $100 million for the DWSRF provided to assist Flint, MI). Division G, Title IV (General Provisions), Section 430, included an additional $600 million ($300.0 million each) within the STAG account for both SRF programs.

P.L. 115-141 also provided $63 million for the WIFIA program, more than doubling the FY2017 appropriation. The act provided $20 million for Alaska Native Village projects and $10 million U.S.-Mexico border projects. It also provided $1.08 billion for state categorical grants, which support a range of environmental programs.

In addition, the act provided the first appropriations for three programs authorized in the WIIN Act (P.L. 114-322, Title II, the Water and Waste Act of 2016): $10 million to help public water systems serving small or disadvantaged communities meet SDWA requirements; $20 million to support lead reduction projects, including lead service line replacement; and $20 million to establish a voluntary program for testing for lead in drinking water at schools and child care programs.[37]

FY2019

The Trump Administration's FY2019 budget request proposed $6.15 billion overall for EPA. The request included $1.394 billion for clean water SRF capitalization grants and $863 million for drinking water SRF capitalization grants (the same amounts *requested* in FY2018). The request included $20 million for the WIFIA program: $17 million to cover subsidy costs, which EPA estimated would allow the agency to lend approximately

---

[37] For more information, see CRS In Focus IF10883, *Overview of U.S. Environmental Protection Agency (EPA) Water Infrastructure Programs and FY2018 Appropriations*, by Mary Tiemann and Jonathan L. Ramseur.

$2 billion (EPA Budget Justification), and $3 million for administrative costs.

In addition, the request proposed $597 million for state categorical grants and $3 million for Alaska Native Village projects. The request proposed to eliminate funding for nonpoint source grants (CWA Section 319)[38] and reduce grant funding for water pollution control (CWA Section 106).[39] In addition, the President's budget request proposed to eliminate funding and U.S.-Mexico border projects.

ACKNOWLEDGMENTS

This report was originally written by Claudia Copeland, who has retired from CRS.

---

[38] 33 U.S.C. §1329.

[39] 33 U.S.C. §1256.

In: Water Quality
Editors: Edward Saunders et al.
ISBN: 978-1-53614-809-1

Chapter 6

DRINKING WATER: EPA NEEDS TO COLLECT INFORMATION AND CONSISTENTLY CONDUCT ACTIVITIES TO PROTECT UNDERGROUND SOURCES OF DRINKING WATER*

United States Government Accountability Office

WHY GAO DID THIS STUDY

Since the early 2000s, increased oil and gas production has resulted in an increase in wastewater that must be managed properly. The majority of wastewater from oil and gas production is injected into underground wells known as class II wells. These wells are regulated to protect drinking water sources under EPA's UIC class II well program and approved state class II

* This is an edited, reformatted and augmented version of United States Government Accountability Office; Report to Congressional Requesters; Accessible Version. Publication No. GAO-16-281, dated February 2016.

programs. EPA oversees state programs, and EPA regions manage programs in states without approval.

GAO was asked to review EPA's oversight of programs' inspection and enforcement information and activities. This report examines the extent to which EPA has collected inspection and enforcement information and conducted oversight activities needed to assess that class II programs protect underground sources of drinking water. GAO reviewed federal and state laws and regulations and EPA guidance and analyzed a nongeneralizable sample of significant violations. GAO interviewed EPA and state officials from programs in a nongeneralizable sample of eight states selected based on shale oil and gas regions, among other factors.

What GAO Recommends

GAO recommends that, among other things, EPA require programs to report well-specific inspections data, clarify guidance on enforcement data reporting, and analyze the resources needed to oversee programs. EPA generally agreed with GAO's findings, but does not plan to require well-specific data and analyze needed resources. GAO continues to believe that EPA should take both actions to better assess if programs protect underground sources of drinking water.

What GAO Found

The Environmental Protection Agency (EPA) has not collected specific inspection and complete or consistent enforcement information, or consistently conducted oversight activities, to assess whether state and EPA-managed Underground Injection Control (UIC) class II programs are protecting underground sources of drinking water. EPA guidance calls for states and EPA regions to report certain information and for EPA to assess whether programs are effectively protecting underground sources of drinking water, but the agency does not. Specifically:

- EPA annually collects summary data from state and EPA-managed programs on the types of inspections they conduct. However, these data are not specific enough to determine the number of different types of inspections that states and EPA regions are to conduct to meet their annual goals. Such goals are specified at the well level (e.g., to inspect 100 percent of wells associated with emergency responses). Under federal internal control standards, managers are to compare actual performance to planned or expected results and analyze significant differences. Without well-specific data on inspections, EPA cannot assess whether state and EPA-managed programs are meeting annual inspection goals.
- EPA collects information on unresolved significant violations of state and EPA-managed programs to determine if the agency needs to take action to enforce applicable program requirements. However, GAO's analysis of a nongeneralizable sample of 93 significant violations for fiscal years 2008 through 2013 found that state and EPA-managed programs did not report data on such violations completely or consistently. For example, of 29 such violations that had not been enforced after 90 days as required, programs reported 7 to EPA. According to EPA and state officials, the cause was inconsistent interpretations of EPA's reporting guidance. EPA officials said they are aware that the data reported on such violations are not complete or consistent, but the agency has not clarified in guidance what data programs should report. Until it does so, EPA does not have reasonable assurance that it has the data needed to assess if it must take enforcement action.

EPA has not consistently conducted oversight activities necessary to assess whether state and EPA-managed programs are protecting underground sources of drinking water. For example, GAO found in June 2014 that EPA does not consistently conduct oversight activities, such as annual on-site program evaluations. According to EPA guidance, such evaluations should include a review of permitting and inspection files or activities to assess whether the state is protecting underground water. In California, for example, EPA did not regularly review permitting, and in July 2014, after a state review of permitting, EPA determined that the program was out of compliance with state and EPA requirements. EPA officials said that they have few resources to oversee UIC class II programs, but EPA has not conducted a workforce analysis consistent with GAO's work on strategic human capital management to identify the resources needed for such oversight. Without conducting such an analysis, EPA will not be able identify the human capital or other resources needed to carry out oversight of the UIC class II programs to help ensure that they protect underground sources of drinking water.

ABBREVIATIONS

Division	California Division of Oil, Gas, and Geothermal Resources
EPA	Environmental Protection Agency
Board	California State Water Resources Control Board
MOA	memorandum of agreement
UIC	Underground Injection Control

441 G St. N.W.
Washington, DC 20548

February 26, 2016

Congressional Requesters

Since the early 2000s, increased oil and gas production across the nation has resulted in a corresponding increase in wastewater that must be managed, reused, or disposed of properly, according to a June 2015

Environmental Protection Agency (EPA) draft report.[1] According to the same report, the growth in domestic oil and gas production has also raised concerns about potential effects to human health and the environment, including the potential contamination of underground drinking water sources by injecting wastewater associated with the production of oil and gas. In particular, the hydraulic fracturing process—in which water, sand, and chemical additives are injected under high pressure to create and maintain fractures in underground formations—allows oil and natural gas from unconventional sources, such as shales, tight sandstones, and coalbed formations, to be developed and can produce large volumes of wastewater. Underground wells used for injecting wastewater and other fluids associated with the extraction of oil and gas resources are known as class II injection wells.[2] The fluids injected into class II wells are composed mostly of saltwater and may contain pollutants such as chlorides, hydrocarbons, and naturally occurring radioactive materials originating from geologic formations containing oil and gas.[3]

EPA and states regulate three types of class II wells associated with oil and gas production: (1) enhanced recovery wells into which brine, water, steam, carbon dioxide, or other fluids and gases are injected into oil- or gas-bearing formations to increase the recovery of residual oil and gas; (2) disposal wells into which brines and other fluids brought to the surface during oil and gas production activities are injected for disposal; and (3) storage wells into which liquid petroleum products are injected, generally as part of the U.S. Strategic Petroleum Reserve.[4] As of 2013, there were over 176,000 class II injection wells in the United States, located in states

[1] EPA, *Draft Assessment of the Potential Impacts of Hydraulic Fracturing for Oil and Gas on Drinking Water Resources* (Washington, D.C.: June 2015).

[2] EPA regulates six classes of underground injection wells.

[3] Fluids that are injected underground for disposal or to enhance recovery are regulated under the Safe Drinking Water Act. While hydraulic fracturing involves the injection of fluids underground for production purposes, the Energy Policy Act of 2005 exempted the process of injecting fluids—other than diesel fuel—into a well to hydraulically fracture formations. However, water that is produced from formations during oil and gas production, including water from hydraulic fracturing activities that flows back out of the well, needs to be disposed of or reused.

[4] The U.S. Strategic Petroleum Reserve is an emergency stock of oil maintained by the U.S. Department of Energy.

as geographically dispersed as California, New Mexico, Oklahoma, Pennsylvania, Texas, and Virginia. As the volume of wastewater generated from oil and gas production has increased, the demand for underground injection wells for disposal of fluids has increased, particularly in states with limited injection capacity, such as Pennsylvania. In addition, some operators have violated federal regulations and disposed of fluids illegally, such as a Kentucky company that in 2013 pled guilty to illegally injecting fluids into sinkholes and an unpermitted well.

To protect underground sources of drinking water, class II injection wells are subject to regulation by the Underground Injection Control (UIC) program overseen by EPA under the Safe Drinking Water Act. The act includes provisions for states to request approval from EPA to manage the class II program in their respective state boundaries. States with approved programs have primary responsibility, or primacy, for managing and enforcing their programs, and EPA has responsibility for managing and enforcing programs in states without primacy. We refer to programs managed by states as state programs and those managed by EPA regional offices as EPA-managed programs. EPA has approved a total of 40 states (26 with class II wells) to manage their programs.[5] Many of these states were approved by EPA in the 1980s, soon after the program's inception. An additional 10 states (7 with class II injection wells) do not have program approval, and the programs in these states are managed by five EPA regional offices.[6] (See app. I for a list of state programs and EPA-managed programs).

The Safe Drinking Water Act establishes general minimum requirements for programs in each state. Applicants for a permit for underground injection must satisfy the state that the injection will not

---

[5] These states regulate over 95 percent of the class II wells nationwide.

[6] According to EPA, the agency is also responsible for management of class II wells on all tribal lands except the Navajo Nation and the Fort Peck Assiniboine and Sioux Tribes. This report only discusses state programs and EPA-managed programs in states and does not directly address tribal or territorial programs. In our June 2014 report, we reported that EPA had approved 39 states for primacy. Tennessee received approval to manage its class II program in 2015, after we issued our report *Drinking Water: EPA Program to Protect Underground Sources from Injection of Fluids Associated With Oil and Gas Production Needs Improvement*, GAO-14-555 (Washington, D.C.: June 27, 2014).

endanger drinking water sources, among other requirements. EPA and states rely on more detailed regulatory and policy safeguards to prevent fluids from migrating into aquifers that can be used as underground sources of drinking water. These safeguards require well operators to meet technical standards for constructing, operating, testing, and monitoring injection wells, among other requirements. In addition, if certain conditions are met, aquifers can be exempted from protection under the act and used for injection. An aquifer may be exempted if (1) it does not currently serve as a source of drinking water and (2) it will not in the future serve as a source of drinking water. If the state has primacy, well operators may request an exemption for injecting fluids into a particular aquifer or portion of an aquifer. The state must submit a request for the aquifer exemption to EPA for review and approval, and if EPA approves, operators may be permitted to inject fluids into the aquifer.

Given the increase in oil and gas production and wastewater disposal, EPA's oversight and enforcement of the UIC class II program is important to ensure that state and EPA-managed programs are protecting underground sources of drinking water. In June 2014, we reported on EPA and state roles, responsibilities, and resources for managing the class II program; EPA and selected state safeguards to protect underground sources of drinking water; EPA's oversight and enforcement of class II programs; and the reliability of data to report on the class II program nationwide.[7] In our June 2014 report, we did not review EPA oversight of state and EPA-managed inspections and enforcement information and activities, which you requested that we review.

This report examines EPA's UIC class II program to determine the extent to which EPA has collected the inspection and enforcement information needed, and conducted the oversight activities necessary, to assess that state and EPA-managed programs are protecting underground sources of drinking water.

To perform this work, we reviewed and analyzed the Safe Drinking Water Act and EPA regulations and guidance on the UIC class II program.

---

[7] GAO-14-555.

We interviewed EPA UIC program officials in the eight regional offices with the highest number of class II wells. To understand the class II program at the state level, we interviewed state officials and reviewed state program documentation for the same sample of states from our June 2014 report on the UIC program.[8] Specifically, we selected a nongeneralizable sample of eight states with class II programs.[9] Two of these states are managed by EPA regions—Kentucky and Pennsylvania—and the remaining six—California, Colorado, North Dakota, Ohio, Oklahoma, and Texas—are state programs. We selected these states from the six shale oil and gas regions defined by the Energy Information Administration.[10] For each of the six oil and gas shale regions, we selected at least one state that had among the highest number of class II injection wells. In July 2014, after we issued our June 2014 report and before we started the work on this review, EPA determined that one of the state programs in the eight states we reviewed, California's class II program, was not in compliance with state and EPA requirements. EPA Region 9 officials and California's UIC program officials have since agreed to a plan to improve the California program over the next several years. We interviewed EPA headquarters, EPA Region 9, and California officials regarding the deficiencies in California's program, the agreed-upon improvement plan, and EPA oversight of California's progress. A summary of the deficiencies found by EPA and California, and California's plans to improve its program, can be found in appendix II. Because EPA determined in July 2014 that California's program was not in compliance with state and EPA requirements, we chose not to include California in our detailed analysis of inspection and enforcement information from the states. Thus, the results of our review of inspection and enforcement information reflect the seven remaining states in our sample.

---

[8] GAO-14-555.

[9] Because the sample is a nongeneralizable sample, our results cannot be generalized to other states but do provide detailed examples of EPA's and states' management of class II programs.

[10] The Energy Information Administration is a statistical agency within the Department of Energy that provides independent data, forecasts, and analyses on energy. Energy Information Administration, *Review of Emerging Resources: U.S. Shale Oil and Shale Gas Plays* (July 2011).

To analyze the information EPA needs and oversight activities necessary to assess whether state and EPA-managed programs are protecting underground sources of drinking water, particularly inspection and enforcement information, we reviewed EPA regulations and guidance and obtained and analyzed data collected by EPA from the states and regions on forms called 7520-3 and 7520-4 forms for fiscal years 2008 through 2013, the most recent years of data available when we began our audit work. We interviewed EPA officials about the collection of these data and determined that they were sufficiently reliable for purposes of reporting data for individual states. To determine if EPA collects information it needs on inspections, we used EPA's 1987 Underground Injection Control Program Compliance Strategy for Primacy and Direct Implementation Jurisdictions (Strategy),[11] which lays out minimum inspection goals that state and EPA-managed programs should set. We then analyzed inspection data from 7520-3 forms for fiscal year 2013, the most current data available, for the seven states to determine the extent to which they provide information to assess inspection goals set by states and EPA regions.

To determine if EPA collects information it needs on enforcement actions, we used EPA's Strategy, which directs states and EPA-managed programs to ensure that timely and appropriate enforcement actions are taken, and EPA's 1986 Reporting Requirements—Underground Injection Control Program Guidance, which provides guidance on what information on violations and enforcement actions should be reported by state and EPA-managed programs on 7520-4 forms.[12] We then selected a nongeneralizable sample of 134 notices of violation, from fiscal years 2008 through 2013 from the seven state and EPA-managed programs we reviewed and compared the information collected on each violation and any related enforcement action to the enforcement data provided to EPA

[11] EPA, Office of Drinking Water, *Underground Injection Control Program Compliance Strategy for Primacy and Direct Implementation Jurisdictions* (Washington, D.C.: Mar. 31, 1987).

[12] EPA, Office of Drinking Water, *Underground Injection Control Program Compliance Strategy for Primacy and Direct Implementation Jurisdictions*. EPA, Office of Drinking Water, *FY 1987 Reporting Requirements—Underground Injection Control Program Guidance (UICP) Guidance #53* (Washington, D.C.: December 1986).

on the 7520-4 forms for that period.[13] We selected a nongeneralizable sample of at least six notices of violation in each of the seven states in our sample. We selected violations on the basis of their significance,[14] the type of enforcement action taken, and the number of days from when the operator was notified to when the violation was resolved, termed returning to compliance with EPA and state requirements. We analyzed the number of days that each of 93 significant violations (of 134) in our sample had been open and compared this to the number of days (90) established by EPA as timely for resolving significant violations; we also analyzed each significant violation to determine if an appropriate enforcement action was taken. The Strategy specifies the appropriate enforcement action for significant violations as a formal enforcement action, which among other things, is legally enforceable, explicitly requires the well owner to take corrective action, and specifies a timetable for completion.[15]

Finally, to assess the oversight activities that EPA has conducted to ensure that programs are protecting underground sources of drinking water, we reviewed findings and recommendations from our June 2014 report and interviewed EPA officials on what the agency has done to implement our recommendations. In addition to the two oversight activities needed to manage UIC class II programs discussed in our June 2014 report, we interviewed EPA officials about another oversight activity—developing and maintaining a database on aquifer exemptions.

---

[13] For the purposes of this report, we refer to all written notifications to operators that they are in violation of state or EPA requirements as notices of violation. During our review of seven states, we noted that state or EPA-managed programs may initiate a single enforcement case against an operator for multiple violations, and so a single notice of violation may cover more than one violation.

[14] Significant violations include those that pose a significant danger to underground sources of drinking water, according to EPA. Significant violations include well operation without mechanical integrity, which causes the movement of fluid outside the authorized zone; well operation at an injection pressure that exceeds the permitted or authorized injection pressure and causes the movement of fluid outside the authorized zone of injection; or the plugging and abandonment of an injection well in an unauthorized manner.

[15] According to the *Strategy*, to take timely and appropriate enforcement actions against well operators with significant violations, programs should take one of the following actions within 90 days after a significant violation is identified: (1) verify that the well operator has returned to compliance, (2) place the well operator on an enforceable compliance schedule and track to ensure future compliance, or (3) initiate a formal enforcement action against the well operator. If the significant violation is not returned to compliance or addressed with a formal enforcement action, it should be reported on the 7520-4 after 90 days.

We conducted this performance audit from October 2014 to February 2016 in accordance with generally accepted government auditing standards. Those standards require that we plan and perform the audit to obtain sufficient, appropriate evidence to provide a reasonable basis for our findings and conclusions based on our audit objectives. We believe that the evidence obtained provides a reasonable basis for our findings and conclusions based on our audit objectives. A more detailed description of our objectives, scope, and methodology is presented in appendix III.

BACKGROUND

This section presents information on roles and responsibilities of EPA and states in the UIC class II program, information on UIC class II inspection and enforcement processes, information collected from state and EPA-managed programs, and activities to oversee state and EPA-managed programs.

EPA and State Roles and Responsibilities for Class II UIC Programs

The UIC class II program is overseen by EPA headquarters and managed by states or EPA regions, depending on whether the state has received primacy. States can obtain primacy in one of two ways. Under section 1422 of the Safe Drinking Water Act, a state can adopt and implement a program that meets specific requirements established under EPA regulations and conduct reporting as EPA requires. Alternatively, under section 1425 of the act, a state can seek approval to manage its own program by demonstrating to EPA that the program is effective in preventing the contamination of underground sources of drinking water. Both types of program must meet four key requirements in the act: (1) they must prohibit unauthorized injections; (2) authorized injections must not endanger drinking water sources; (3) they must include inspection, monitoring, recordkeeping, and reporting requirements; and (4) they must apply their provisions to federal agencies and federal land.[16] However, states approved by this alternative process do not need to address all of the specific requirements, such as those related to well construction and testing, established in EPA regulations.

Program oversight by EPA headquarters includes issuing regulations and guidance, assessing implementation of regulations and guidance by state and EPA-managed programs, and gathering information and reporting it. EPA regions both oversee state programs that have primacy and manage programs in states that do not have primacy, and states with primacy manage their own programs. Management includes permitting wells; inspecting wells; enforcing regulations and implementing guidance; reporting information on well inventories, inspections, violations, and enforcement actions; and investigating instances of potential contamination of aquifers.

EPA issued a series of guidance documents describing the program and various responsibilities of states and EPA regions. To oversee state and EPA-managed programs and to ensure that they are protecting underground sources of drinking water, EPA collects certain information and conducts certain activities, as described in several guidance documents. Specifically:

- ***Memorandum of agreement for the UIC program***. Issued in 1981, this guidance directs EPA regions to enter into a memorandum of agreement (MOA) with each primacy state that includes the terms, conditions, or agreements between the state and EPA regarding the administration and enforcement of state program requirements, [17] including state inspection, enforcement, and reporting requirements.[18]

---

[16] 42 U.S.C. § 300h(b)(1) (2015).

[17] For example, 40 C.F.R. § 145.25(b) requires the MOA with states that received primacy under section 1422 of the act (1422 states) to include (1) provisions for the prompt transfer from EPA to the state of pending permit applications and any other information relevant to program operation not already in the possession of the state director (e.g., support files for permit issuance and compliance reports); (2) provisions specifying classes and categories of permit

- ***Reporting Requirements—Underground Injection Control Program (Program Reporting).*** Issued in 1986, this guidance explains and clarifies the information state and EPA-managed programs are to report.[19]
- ***Underground Injection Control Program Compliance Strategy for Primacy and Direct Implementation Jurisdictions (Strategy).*** Issued in 1987, this document provides guidance to state and EPA-managed programs on well inspections and enforcement of program requirements, including information that should be reported on inspections and enforcement actions.[20]
- ***Interim Guidance for Overview of the Underground Injection Control Program (Program Oversight).*** Issued in 1983, this document provides guidance to EPA regions and headquarters on activities that they should take to effectively oversee state and EPA-managed programs, respectively.[21]
- ***Guidance for Review and Approval of State Underground Injection Control Programs and Revisions to Approved State Programs.*** Issued in 1984 to provide guidance for EPA regions on the review and approval of changes to state program

---

applications, draft permits, and proposed permits that the state will send to the regional administrator for review, comment, and, where applicable, objection; (3) provisions specifying the frequency and content of reports, documents, and other information that the state is required to submit to EPA (the state shall allow EPA to routinely review state records, reports, and files relevant to the administration and enforcement of the approved program, and state reports may be combined with grant reports where appropriate); (4) provisions on the state's compliance monitoring and enforcement program; (5) when appropriate, provisions for joint processing of permits by the state and EPA, for facilities or activities that require permits from both EPA and the state under different programs; and (6) provisions for modification of the MOA.

[18] EPA, Office of Drinking Water, *Memorandum of Agreement (MOA) for the Underground Injection Control Program (UIC) Ground-Water Program Guidance #14* (Washington, D.C.: May 1981).

[19] EPA, Office of Drinking Water, *FY 1987 Reporting Requirements—Underground Injection Control Program Guidance (UICP) Guidance #53*.

[20] EPA, Office of Drinking Water, *Underground Injection Control Program Compliance Strategy for Primacy and Direct Implementation Jurisdictions*.

[21] EPA, Office of Drinking Water, *Interim Guidance for Overview of the Underground Injection Control (UIC) Program, Ground-Water Program Guidance #30* (Washington, D.C.: June 1983).

requirements,[22] this document includes guidance for EPA regions and headquarters on how to review and approve requests to exempt aquifers and how decisions on aquifer exemptions should be documented and reported.

- ***Enhancing Coordination and Communication with States on Review and Approval of Aquifer Exemption Requests under the Safe Drinking Water Act (Aquifer Exemption Coordination).*** Issued in 2014, this document provides guidance on how to improve coordination and recordkeeping on aquifer exemption decisions among states, EPA regions, and EPA headquarters.[23]

According to EPA's 1981 MOA guidance, EPA regions should develop an MOA with each primacy state to outline areas of the applicable regulations that are relevant to the administration and enforcement of the state's program requirements, including clarifying

- EPA and state roles and responsibilities and the process for sharing information between EPA and the state;
- state responsibilities for expeditiously drafting, circulating, issuing, modifying, reissuing, and terminating permits, consistent with applicable regulations;
- state responsibilities to operate a timely and effective system to track well operator compliance with program requirements, including inspection procedures;
- state responsibilities for taking timely and appropriate enforcement action against persons in violation of program requirements, including use of effective enforcement tools such as penalties; and

[22] EPA, Office of Drinking Water, *Guidance for Review and Approval of State Underground Injection Control (UIC) Programs and Revisions to Approved State Programs GWPB Guidance #34* (Washington, D.C.: January 1984).

[23] EPA, Office of Ground Water and Drinking Water, *Enhancing Coordination and Communication with States on Review and Approval of Aquifer Exemption Requests Under the Safe Drinking Water Act* (Washington, D.C.: July 14, 2014).

- state reporting requirements, including the type and frequency of data to be reported, and EPA's annual evaluation of the state program.[24]

According to EPA's 1983 Program Oversight guidance, EPA-managed programs are also responsible for establishing systems to track well operator compliance; taking timely and appropriate action to resolve violations, including use of effective enforcement tools; and reporting data on the program.[25]

Information on the UIC Class II Program Inspection and Enforcement Process

Injection well inspections, to discover and deter violations, and enforcement are identified in EPA's 1987 Strategy as tools to achieve operator compliance with applicable requirements.[26] According to the Strategy, each state and EPA-managed program should have a strategy for identifying how many wells it should inspect and the types of inspections to be conducted at these wells. The types of inspections that state and EPA inspectors conduct can vary from routine inspections that ensure that well sites are being properly maintained, to inspections that include observing pressure tests to determine if wells are structurally sound, known as mechanical integrity tests (see app. IV for information on types of inspections).

The enforcement process begins once an inspector identifies a violation. Violations of UIC program requirements can involve a number of actions on the part of well operators, such as injecting fluids without authorization, injecting fluids at pressures above those permitted, or failing

---

[24] EPA, Office of Drinking Water, *Memorandum of Agreement (MOA) for the Underground Injection Control Program (UIC) Ground-Water Program Guidance #14*.

[25] EPA, Office of Drinking Water, *Interim Guidance for Overview of the Underground Injection Control (UIC) Program, Ground-Water Program Guidance #30*.

[26] EPA, Office of Drinking Water, *Underground Injection Control Program Compliance Strategy for Primacy and Direct Implementation Jurisdictions*.

to show that a well holds pressure during testing (mechanical integrity testing). According to EPA's Strategy, a state or EPA-managed program can take various enforcement actions when it finds wells that are violating program requirements. When inspectors identify wells that are violating applicable program requirements, they typically first notify the well operators of the violations. According to EPA's Strategy, these notifications can be done through discussion or in writing. For more serious violations, state and EPA-managed programs can take stronger enforcement actions.

According to EPA's Strategy, state and EPA-managed programs are to escalate their enforcement response as needed to resolve violations, although the actions taken by a program may depend on a number of factors, including the severity of the violation and its potential to contaminate drinking water sources. Actions to gain compliance with program requirements can include sanctions, such as shutting down a well, assessing administrative penalties, or referring the matter for civil or criminal adjudication (see app. V for details of the enforcement process).

Information Collected from State and EPA-Managed Programs

EPA's regulations and 1986 Program Reporting guidance direct state and EPA-managed programs to report specific information on class II wells to assist with program oversight.[27] The Program Reporting guidance directs these programs to report data on inspections, violations, and enforcement actions. Specifically, the agency collects information from programs on different 7520 forms submitted by state and EPA-managed programs. According to the Program Reporting guidance, information on inspections conducted by state and EPA-managed programs is collected on 7520-3 forms and includes information on the total number of different types of inspections. According to this guidance, EPA also collects information on the number of significant violations and enforcement

[27] EPA, Office of Drinking Water, *FY 1987 Reporting Requirements—Underground Injection Control Program Guidance (UICP) Guidance #53*.

actions conducted by state and EPA-managed programs on 7520-4 forms. Under the Safe Drinking Water Act, EPA is required to notify primacy states of any violations of state UIC programs it discovers and, if a state does not take appropriate enforcement action within 30 days, issue an order or initiate legal action itself.[28] According to EPA guidance, the 7520-4 forms collect information on individual significant violations that threaten underground sources of drinking water to help EPA determine whether it should intervene to enforce state or EPA requirements.

In June 2014, we found that the data on violations and contamination of underground sources of drinking water that EPA collects from its 7520 forms were not sufficiently complete or comparable to allow EPA to aggregate state information and report on the status of the class II program nationally.[29] We also found that EPA was developing a national UIC database to collect comparable, well-specific data from states, but that, as of January 2014, the database was not fully populated. We recommended in our June 2014 report that to support nationwide reporting goals until the national UIC database is complete, EPA develop and implement a protocol for states and regions to enter 7520 data consistently and for regions to check 7520 data for consistency and completeness to ensure that data collected from state and EPA-managed class II programs are complete and comparable for purposes of reporting at a national level. EPA agreed that there is room for improvement in the completeness and consistency of data submitted by the states and regions through the 7520 forms. In response to our recommendation, according to EPA officials, the agency has proposed updated 7520 instructions, intended to encourage consistent reporting by states and regions. The updated instructions have not been finalized and, according to EPA officials, cannot be used for reporting until they are approved by the Office of Management and Budget.[30] EPA has also developed new standard operating procedures that update protocols for EPA regional review of 7520 reports submitted by state programs and

---

[28] 42 U.S.C. §1423(a)(1) (2015).

[29] GAO-14-555. This refers to data from EPA's 7520-2 form, which collects violation information.

[30] According to EPA officials, as of December 2015, the Office of Management and Budget was reviewing the updated instructions.

headquarters review of 7520 reports submitted by EPA-managed programs.

Activities to Oversee State and EPA-Managed Programs

EPA's regulations and 1983 Program Oversight guidance[31] direct EPA headquarters and regions to conduct specific activities to ensure that the state and EPA-managed programs are protecting underground sources of drinking water. These activities include conducting annual on-site evaluations of state and EPA-managed programs. In addition, EPA regulations require the agency to incorporate state program requirements, and any changes to them, into federal regulations to be able to enforce them if necessary, and to approve aquifers for exemption from protection under the act, as appropriate, to allow injection of fluids.

According to EPA's Program Oversight guidance, EPA regional officials are expected to conduct annual on-site evaluations of state programs.[32] These evaluations involve, among other things, an on-site meeting with state UIC officials to discuss program performance and can include a review of inspection and enforcement files, both of which are intended to help determine whether the state program is effective at protecting underground sources of drinking water. We found in June 2014, however, that EPA was not consistently carrying out annual on-site evaluations of state class II programs.[33] According to EPA officials at the time, limited resources have prevented EPA regions and headquarters from consistently conducting on-site reviews, and some of the oversight activities identified in the Program Oversight guidance may no longer be needed. We recommended, and EPA agreed, that EPA should evaluate, and revise as needed, UIC program guidance on effective oversight to identify

---

[31] EPA, Office of Drinking Water, *Interim Guidance for Overview of the Underground Injection Control (UIC) Program, Ground-Water Program Guidance #30.*

[32] EPA, Office of Drinking Water, *Interim Guidance for Overview of the Underground Injection Control (UIC) Program, Ground-Water Program Guidance #30.*

[33] GAO-14-555.

essential activities that EPA headquarters and regions need to conduct to effectively oversee state and EPA-managed programs.

According to EPA regulations, EPA is also required to incorporate state program requirements and changes to those requirements into federal regulations. Under its regulations, EPA can only enforce state program requirements that it has incorporated into federal regulations. In June 2014, we found that EPA was not consistently incorporating state program requirements, or changes to state program requirements, into federal regulations, and as a result, EPA had not been able to enforce at least one state's program requirements.[34] To ensure that EPA maintains enforcement authority of state program requirements, we recommended that EPA conduct a rulemaking to incorporate state program requirements, and changes to state program requirements, into federal regulations and, at the same time, evaluate and consider alternative processes to more efficiently incorporate future changes to state program requirements into federal regulations without a rulemaking.[35] EPA disagreed with this recommendation and said that in lieu of a single rulemaking, it was conducting an ongoing process of individual rulemakings to approve and codify state program revisions, as discussed later in the report.

According to EPA regulations and EPA's 2014 Aquifer Exemption Coordination memorandum, EPA is responsible for the final review and approval of all aquifer exemption requests. Well operators seeking an aquifer exemption to conduct injection activities in a state with primacy typically submit the exemption application to state program officials along with supporting information. State program officials are to review the application and, if the information submitted supports an exemption, submit a request to approve the exemption to the appropriate EPA regional office. Applicants in states with EPA-managed programs are to submit applications directly to the EPA region managing the program, and the

---

[34] GAO-14-555.

[35] The Safe Drinking Water Act requires that state UIC program revisions made in response to changes in EPA UIC regulations be approved by rule. Our recommendation focused on the requirement in EPA's regulation, but not in the Safe Drinking Water Act, that effectively prohibits EPA enforcement of state UIC regulations unless these regulations are codified in the federal regulations.

region approves or disapproves the exemption applications. EPA regions are responsible for maintaining documentation supporting the decision to exempt an aquifer and a record of all exempted aquifers. According to the Aquifer Exemption Coordination memorandum, maintaining the decision memos and records underlying EPA's approval or disapproval of exemption applications and standardized, readily available data on all existing aquifer exemptions is important to supporting informed decisions about uses for drinking water.

Under the act, if EPA determines that a state program is no longer protecting underground sources of drinking water, the agency can revoke a state's primacy by rule. According to EPA officials, before such a point is reached, the agency can work with the state to return the state's program to compliance with EPA and state UIC class II regulations. For example, in July 2014, after California identified instances in which it had authorized injection into nonexempt aquifers, EPA determined that the state's program was not in compliance with state and EPA requirements. In a series of letters from July 2014 through July 2015, EPA and the state's Division of Oil, Gas, and Geothermal Resources reached agreement on a plan to improve California's program. (See app. II for the details of the status of California's program).

EPA Has Not Collected Information, or Consistently Conducted Activities, Necessary to Assess if Programs Are Protecting Underground Sources of Drinking Water

EPA has not collected inspection and enforcement information, or consistently conducted specific oversight activities, to assess whether state and EPA-managed programs are protecting underground sources of drinking water. EPA's 1981 MOA guidance directs states and EPA regions to include provisions in memorandums with states to ensure that regional offices can collect the information and conduct the activities necessary for oversight, including (1) collecting information on inspections and enforcement actions and (2) conducting activities to incorporate approved changes to state program regulations into federal regulations, conducting annual on-site program evaluations, and reviewing and approving aquifer exemption applications.[36] EPA's Program Oversight guidance also states that EPA headquarters should collect the same information and conduct the same activities to oversee programs managed by EPA regions where applicable.

EPA Has Not Collected Inspection and Enforcement Information to Assess Whether Programs Are Protecting Underground Sources of Drinking Water

EPA has not collected inspection and enforcement information that can be used to assess whether state and EPA-managed programs are effectively protecting underground sources of drinking water.

Inspections

EPA collects information from state and EPA-managed programs on the types of inspections they conduct, but the information EPA collects is at a summary level and not specific enough to assess whether states are meeting inspection goals established to protect underground sources of drinking water.[37] In the 1987 Strategy, EPA provides guidance on the types of UIC inspections that state and EPA-managed programs should conduct and specifies minimum annual inspection goals (i.e., frequency of each inspection type) for state and EPA-managed programs. For example, (1) 100 percent of wells associated with emergency responses and public complaints should be inspected annually,[38] (2) 25 percent of mechanical

---

[36] EPA, Office of Drinking Water, *Memorandum of Agreement (MOA) for the Underground Injection Control Program (UIC) Ground-Water Program Guidance #14*.

[37] The Safe Drinking Water Act requires class II programs to have inspection requirements to protect underground sources of drinking water.

[38] According to the *Strategy*, public complaints warrant prompt attention but should be evaluated (e.g., with a telephone call) to determine their veracity.

integrity tests conducted annually should be witnessed by an inspector, and (3) routine inspections to verify that wells are operating in compliance with applicable requirements should be conducted at least once every 5 years. According to the Strategy, state and EPA-managed programs should set goals for different types of inspections based on factors such as available resources and program priorities (see app. IV for additional information on EPA guidance on inspections and selected state inspection programs).

EPA's 1987 Program Reporting guidance states that the inspection data that EPA collects from state and EPA-managed programs should be used to track each program's progress toward meeting its inspection goals, which are to be based on EPA's minimum annual inspection goals.[39] EPA's minimum annual inspection goals are specified at the well level (e.g., 100 percent of wells associated with emergency responses). However, state and EPA-managed programs report annual summary data on the number of inspections conducted for each inspection type by state and not data on which wells were inspected, when they were inspected, the types of inspection conducted at each well, and the results of those inspections. For example, the summary data EPA collects on routine inspections, as shown in *Table 1*, could not be used to determine if a state or EPA-managed program had conducted a routine inspection of each of its class II wells over a 5-year period or multiple inspections of individual wells. For the seven state and EPA-managed programs we reviewed, annual data reported to EPA included the total number of wells inspected and types of inspections conducted statewide, as shown in *Table 1* for fiscal year 2013.

Because the inspection data that EPA has collected from states have not been well-specific and therefore have not included the total number of inspections by type that could have been done, EPA's ability to track each state program's progress toward meeting its inspection goals is limited.

---

[39] EPA, Office of Drinking Water, *FY 1987 Reporting Requirements—Underground Injection Control Program Guidance (UICP) Guidance #53*.

Table 1. Inspections Reported by Selected State and EPA-Managed Programs for Fiscal Year 2013

State	Emergency and complaint response	Mechanical integrity tests witnessed	Well pluggings witnessed	Well constructions witnessed	Routine inspections
Colorado	0	275	4	6	601
Kentucky	9	213	10	0	871
North Dakota	7	343	9	75	8,474
Ohio	0	57	3	18	2,010
Oklahoma	Not reported	3,368	Not reported	Not reported	6,378
Pennsylvania	0	245	9	16	42
Texas[a]	475	6,601	Not reported	Not reported	23,242

Legend: EPA = Environmental Protection Agency.

Source: GAO analysis of EPA and selected state data. | GAO-16-281.

Notes: We chose not to include California in our detailed analysis of inspection and enforcement information from the states because EPA determined in July 2014 that California's program was not in compliance with state and EPA requirements. The results of our review of inspection and enforcement information reflect the seven remaining state and EPA-managed programs we reviewed. Kentucky and Pennsylvania are EPA-managed programs. Although each state uses different procedures to collect its data, we found that the data were not comparable across states but that they were sufficiently reliable for reporting on a state-by-state basis.

[a]Texas reports on total well pluggings and well constructions witnessed for both production wells and injection wells but cannot provide numbers specific to class II wells.

Under federal standards for internal control, managers need to compare actual performance to planned or expected results and analyze significant differences.[40] EPA officials told us that they recognize that they cannot verify progress toward meeting state program inspection goals without well-specific data on inspections and have made efforts to collect well-specific data through voluntary programs, but do not require its collection. Starting in 2007, EPA had been working to develop a voluntary national UIC database to provide well-specific data from state and EPA-managed programs; however, according to EPA officials in December 2015, Montana was the only participating state program, and the agency plans to complete the national database with Montana and the seven EPA-managed

[40] GAO, *Standards for Internal Control in the Federal Government*, GAO/AIMD-00-21.3.1 (Washington, D.C.: November 1999).

programs currently participating.[41] EPA officials said that they do not have well-specific information because they do not require it and most state programs have not provided it voluntarily through the national UIC database. However, EPA's MOA guidance says that EPA may request and should be given access to all files necessary for evaluating the administration of the state program.[42] Until EPA requires and collects well-specific data on inspections from state and EPA-managed programs, including the types of inspections conducted at each well, when the inspections were conducted, and the results of the inspections, the agency cannot assess whether the programs are meeting their annual inspection goals to protect underground sources of drinking water. EPA officials said that EPA will also have access to another voluntary database being compiled by the Department of Energy that contains additional data from state programs on injection wells. According to the officials, however, the department's database does not provide well-specific information on inspections either.

Enforcement

EPA has not collected consistent or complete enforcement information that can be used to assess whether state and EPA-managed programs are effectively protecting underground sources of drinking water. To carry out the Safe Drinking Water Act's provision that EPA take action on violations that have not been enforced, EPA's 1987 Strategy directs state and EPA-managed programs to take timely and appropriate enforcement action against significant violations of state or EPA requirements. The Strategy defines a timely and appropriate response taken by a state or EPA-managed program as resolving the violation or initiating a formal enforcement action within 90 days of the identification of the violation. To help ensure that violations are addressed in a timely and appropriate way, EPA's 1987 Strategy and 1986 Program Reporting guidance call for state

[41] According to EPA officials, several other class II programs are working on developing their own data submission capability for the national UIC database but are not currently participating.

[42] EPA, Office of Drinking Water, *Memorandum of Agreement (MOA) for the Underground Injection Control Program (UIC) Ground-Water Program Guidance #14*.

and EPA-managed programs to report information to EPA on significant violations that were not resolved within 90 days of discovery and also did not have a formal enforcement action taken against the well operator. The act requires EPA to enforce state program requirements within 30 days after the agency becomes aware that the state has not taken appropriate enforcement action.

However, our review of data collected by EPA on significant violations demonstrated that EPA's ability to take action may be limited by incomplete and inconsistent enforcement data reported by state and EPA-managed programs. Specifically, our analysis of 93 significant violations for fiscal years 2008 thru 2013 for the seven state and EPA-managed programs we reviewed found that there were 29 that were not resolved within 90 days of operator notification and for which formal action had not been taken within that time.[43] According to the Strategy, each of these violations should have been reported on the 7520-4 form by the state to the appropriate EPA region or by the EPA-managed program to EPA headquarters. However, our analysis of the 7520-4 form data showed that state and EPA-managed programs reported 7 of these 29 violations to the agency. *Table 2* shows the results of our analysis of the 7520-4 forms (see app. V for additional information on our analysis and app. VI for the full list of violations and enforcement actions taken).

According to EPA headquarters, regional, and state officials we interviewed, state and EPA-managed programs used different interpretations of the Strategy and Program Reporting guidance to fill out the forms, resulting in incomplete, and potentially inconsistent, information across the programs. EPA headquarters officials told us that all significant violations that were not resolved within 90 days from the date the violation was discovered should be reported on the 7520-4 form quarterly until they are resolved, regardless of whether the program had

[43] EPA's 1987 *Strategy* calls for state or EPA-managed programs to take action within 90 days of identifying the significant violation. We calculated this time frame using the number of days from the time the operator was notified of the significant violation to when the state or EPA-managed program took formal enforcement action, as defined by the *Strategy*. Using this calculation provides the most conservative estimate of the time that a program took to determine if the violation should have been reported to EPA on the state's appropriate 7520 form.

already initiated enforcement action against the well operator. However, EPA's Strategy and Program Reporting guidance call for programs to report, on the 7520-4 form, information on significant violations that (1) were not resolved within 90 days from the date the violation was discovered and (2) had not had a formal enforcement action taken against the well operator. In addition, according to the Program Reporting guidance, significant violations reported on the 7520-4 form should continue to be reported quarterly on subsequent 7420-4 forms until they are resolved.[44] State and EPA officials we interviewed provided different interpretations of what they were to put on the 7520-4 form, which would result in some programs reporting significant violations and some not. Consistent with EPA's Strategy and Program Reporting guidance, officials we interviewed from Ohio and EPA Region 4 (Kentucky) told us that they only report significant violations on the 7520-4 form that were not resolved within 90 days and for which a formal enforcement action had not been taken against the well operator. However, officials we interviewed from North Dakota, Oklahoma, Texas, and EPA Region 3 (Pennsylvania), told us that they report all unresolved significant violations regardless of whether they have taken a formal enforcement action. According to the officials, the information that ultimately gets reported on the 7520-4 form is based on a quarterly calculation of how long the well has been out of compliance; however, according to the officials, EPA only requires state and EPA-managed programs to submit 7520-4 forms to EPA semiannually. In addition, officials in North Dakota and Oklahoma told us that they only report significant violations once and not in subsequent quarters, even if the violations have not been resolved.

EPA headquarters officials told us they are aware that the information reported by states and EPA regions is not complete or consistent, but they have not clarified, in guidance or otherwise, what information should be reported. EPA headquarters officials told us that regions are responsible for

[44] According to EPA officials, the information that ultimately gets reported on the 7520-4 form is based on a quarterly calculation of how long the well has been out of compliance; however, according to the officials, EPA only requires state and EPA-managed programs to submit 7520-4 forms to EPA semiannually and regions can request that the forms be submitted more frequently if needed.

ensuring that state and EPA-managed programs take timely and appropriate enforcement actions, and that regions generally assess the programs' enforcement response on a case-by-case basis through informal communications with state program staff. The information received on the 7520-4 form, however, is the only documented information reported to EPA regions and headquarters on individual violations that may not have been enforced in a timely or appropriate manner. Until it clarifies guidance on what data should be reported on the 7520-4 form, EPA does not have reasonable assurance that state and EPA-managed programs report complete and consistent information on unresolved significant violations or that it has the information it needs to assess whether it must take enforcement action, as directed under the act, to protect underground sources of drinking water.

Table 2. Summary of Selected State and EPA-Managed Programs' Reporting of Significant Violations on the 7520-4 Forms

State	Significant violations	Violations that should have been reported on the 7520-4	Violations that were reported on the 7520-4	Violations that were not reported on the 7520-4
Colorado	1	0	N/A	N/A
Kentucky	29	7	7	0
North Dakota	0	0	N/A	N/A
Ohio	18	2	0	2
Oklahoma	20	0	N/A	N/A
Pennsylvania	5	3	0	3
Texas	20	17	0	17
Total	93	29	7	22

Legend: EPA = Environmental Protection Agency.

Source: GAO analysis of EPA and selected state data. | GAO-16-281

Notes: We chose not to include California in our detailed analysis of inspection and enforcement information from the states because EPA determined in July 2014 that California's program was not in compliance with state and EPA requirements. The results of our review of inspection and enforcement information reflect the seven remaining state and EPA-managed programs we reviewed. Kentucky and Pennsylvania are EPA-managed programs.

EPA Has Not Consistently Conducted Oversight Activities Necessary to Assess Whether Programs Are Protecting Underground Sources of Drinking Water

EPA has not consistently conducted three oversight activities necessary to assess whether state and EPA-managed programs are protecting underground sources of drinking water, as required by regulations and specified in guidance: (1) incorporation of state program requirements, or changes to state program requirements, into federal regulations; (2) the final review and recordkeeping for all aquifer exemption applications it approves; and (3) annual on-site program evaluations.

State Program Requirements

We found in June 2014 that EPA had not consistently incorporated state program requirements, or changes to state program requirements, into federal regulations, as required by agency regulations.[45] Specifically, if a state does not enforce a requirement against an injection well operator violating state regulations, EPA can take enforcement action if EPA has approved the state regulations being violated and incorporated them into federal regulations, and has met specific procedural requirements.[46] EPA regulations and guidance establish a process for EPA and its regions to review and approve state programs, as well as changes to state programs. Under its regulations, EPA can only enforce state program requirements that it has incorporated into federal regulations through a rulemaking process.[47] Where it has not done so, EPA is not able to enforce state program requirements if needed. In June 2014, we found that EPA had not yet incorporated changes to some state program requirements into federal

[45] GAO-14-555.

[46] EPA must give the state notice; if after 30 days the state has failed to commence appropriate action, EPA is to issue an order or begin a court action.

[47] Rulemaking requires EPA to provide public notice of the proposed regulatory changes, respond to the significant issues raised during the comment period and discuss any changes made to the regulation as a result, and publish the text of the final regulation in the *Federal Register*. The *Federal Register* is the daily publication for rules, proposed rules, and notices of federal agencies.

regulations and therefore did not have the ability to enforce these state program requirements if necessary. We concluded that until it conducts a rulemaking to incorporate the backlog of state program requirements and changes to state program requirements that have been approved, EPA would not be able to enforce some state program requirements, hindering its ability to protect underground sources of drinking water. To ensure that EPA maintained enforcement authority of state program requirements, we recommended that EPA conduct a rulemaking to incorporate state program requirements, and changes to state program requirements, into federal regulations. We also recommended that at the same time, EPA evaluate and consider alternative processes to more efficiently incorporate future changes to state program requirements into federal regulations without a rulemaking.

In comments responding to our June 2014 report, EPA disagreed with our recommendation to conduct a rulemaking and said that a single rulemaking would be impractical because the process would take many years to complete and would still not ensure that all program changes were incorporated into federal regulations, as other states could make changes to their programs during this time.[48] In lieu of a single rulemaking, EPA said in its comments that it was conducting an ongoing process of individual rulemakings to approve and codify state program revisions in collaboration with states, EPA regions, and EPA's Office of Enforcement and Compliance Assurance. However, as stated in our June 2014 report, according to an analysis conducted by EPA in 2010, EPA estimated that it would take 2 to 3 years, dedicated EPA personnel, and $150,000 in outside contractor support to identify, approve, and conduct a single rulemaking to incorporate all state program changes made since 1991 into federal regulations. By EPA's own estimate, the targeted state-by-state approach will take much longer than a single rulemaking and will face greater challenges with states continuing to make changes in the interim, leaving EPA without the ability to enforce state programs to protect underground sources of drinking water if needed. EPA provided no evidence in its

[48] GAO-14-555.

comments that individual rulemakings would be any less costly or any more efficient than the approach it assessed in 2010. As of December 2015, EPA has not taken action to incorporate state program requirements, or changes to state program requirements, into federal regulations.

Aquifer Exemptions

EPA is also responsible for the final review, approval, and recordkeeping for all aquifer exemption applications, but the agency does not have the location or supporting documentation necessary to identify the size and location of all aquifers for which it has approved exemptions from protection under the act. According to EPA's 2014 Aquifer Exemption Coordination guidance, EPA regions need to have complete records documenting support for EPA's approval or disapproval of exemption applications to inform decision making by state and EPA-managed programs on injection well permits.[49] According to EPA officials, regional offices generally maintain the most comprehensive and up-to-date data on aquifer exemption approvals. Since 2003, EPA has worked to compile comprehensive information on aquifer exemptions, including data on the aquifers' sizes and locations. In 2011, EPA determined that its headquarters did not have information on all exempted aquifers and requested that EPA regional offices provide information on all aquifers exempted in their respective regions to help compile a centralized database.[50]

According to EPA officials, the agency has compiled a rudimentary database from regional datasets, paper files supporting aquifer exemption decisions, and hard copies of maps specifying the size and location of exempted aquifers. However, EPA officials said that the database of aquifer exemptions does not include complete information on each exemption listed and that EPA does not have a complete inventory of exemptions. In particular, according to EPA officials, the agency is

---

[49] EPA, Office of Ground Water and Drinking Water, *Enhancing Coordination and Communication with States on Review and Approval of Aquifer Exemption Requests Under the Safe Drinking Water Act.*

[50] EPA, Office of Ground Water and Drinking Water, *Underground Injection Control Program - Aquifer Exemptions - New Regional Reporting Process* (Washington, D.C.: July 2011).

missing information on exemption decisions made when state programs were granted primacy in the 1980s because the supporting documentation is not readily accessible or was damaged while in storage.

If EPA had maintained an updated database on aquifer exemptions, then EPA Region 9 may have had the information it needed to review injection well permits to determine whether injections were being made into exempted aquifers in California. Instead, California discovered that it had authorized injection into nonexempt aquifers. Specifically, EPA requested additional information on aquifer exemptions from California in 2012 as a part of EPA's review of historical data on aquifer exemptions nationwide. At that time, the state reviewed supporting documentation for the aquifer exemptions and the associated injection wells and determined that it had permitted operators to inject into nonexempt aquifers that the state believed were exempted in the 1980s, when EPA granted primacy to California to manage the class II program.

In July 2014, after identifying water supply wells in the vicinity of some of these injection wells and informing EPA Region 9, California ordered operators of those injection wells to cease injection into certain nonexempt aquifers, and to submit data to California so the threat to underground sources of drinking water and human health could be assessed. In July 2014, as a result of this issue, EPA determined that California's program was not in compliance with state and EPA requirements and supported California's plan to review injection wells that were permitted to inject into nonexempt aquifers. As of October 2015, California had identified over 500 wells injecting into 11 nonexempt aquifers with the potential to threaten underground sources of drinking water, and 23 of those wells had been shut-in, or ceased injecting fluids. In November 2015, California shut-in an additional 33 injection wells injecting into nonexempt aquifers. As of October 2015, California officials said that they are continuing to collect information on wells injecting into nonexempt aquifers to determine if additional wells should be shut-in to protect underground sources of drinking water and are working with EPA Region 9 to collect additional information on aquifer exemptions to help complete EPA's database.

As of December 2015, EPA officials told us that the majority of aquifers in its database of approved exemptions have complete size and location data and that headquarters continues to collect information from the regions and state programs to fill in the remaining data gaps and ensure that the database is complete and accurate. The officials told us that for this reason, it is unlikely that they will discover deficiencies in recordkeeping for approved aquifer exemptions similar to those identified in California. However, while EPA officials believe that they have the majority of the data on aquifer exemptions, the database does not include some historical data on exemption decisions made when state programs were granted primacy in the 1980s. In addition, the database only has aquifer exemption data through 2011 and is missing data on aquifer exemptions approved over the past 4 years. According to EPA officials, the database is a headquarters-based spreadsheet and updates with new approvals on aquifer exemptions will need to be collected from EPA regions and entered manually.[51] The officials also said that EPA will complete the database using 2011 data and only plan to add updated data if sufficient resources are available. Until it has a complete aquifer exemption database and a way to update it periodically, EPA does not have sufficient information on aquifer exemptions to oversee state and EPA-managed programs and assess whether programs are protecting underground sources of drinking water.

Annual On-site Program Evaluations

As we reported in June 2014, EPA has not consistently conducted annual on-site program evaluations, as directed by its 1983 Program Oversight guidance.[52] This guidance directs EPA regions and headquarters to conduct annual on-site program evaluations of state and EPA-managed programs, which it characterizes as a key activity necessary for effective

---

[51] According to EPA officials, information collected by headquarters from the regions is contained in two data files: (1) an Excel spreadsheet with data on aquifer exemptions and (2) a geospatial data file that will allow the aquifer exemption boundaries to be displayed on a United States map.

[52] EPA, Office of Drinking Water, *Interim Guidance for Overview of the Underground Injection Control (UIC) Program, Ground-Water Program Guidance #30.*

oversight, and to ensure that state and EPA-managed class II programs protect underground sources of drinking water.[53] According to EPA's Program Oversight guidance, EPA regions should perform at least one on-site evaluation of each state program each year to assess whether the state is managing the program consistent with state regulations, setting program objectives consistent with national and regional program priorities, and implementing recommendations from previous evaluations, among other activities. According to the Program Oversight guidance, annual on-site evaluations of state programs should also include a review of permitting and inspection files or activities to assess whether the state program is protecting underground sources of drinking water. In particular, because permitting files should include information on the well location, and the geology and aquifers in the area surrounding the injection well, a review of permitting files should cover this information. EPA headquarters is responsible for conducting similar on-site program evaluations of EPA-managed programs.

In our June 2014 report, regional officials said that on-site program evaluations are valuable for coordinating between federal and state officials to improve program management. According to EPA officials at the time, however, limited resources have prevented regions, and EPA headquarters, from consistently conducting on-site program evaluations. To ensure effective oversight of the class II program, in June 2014, we recommended, and EPA agreed, that EPA evaluate and revise, as needed, UIC program guidance on effective oversight to identify essential activities that EPA headquarters and regions need to conduct to effectively oversee state and EPA-managed programs to ensure that they were effective at protecting underground sources of drinking water.

If EPA had conducted oversight activities, such as annual on-site program evaluations, EPA Region 9 may have discovered that California's class II program did not comply with state and EPA requirements before 2014. In particular, regular on-site program evaluations that included reviews of permitting files may have identified the deficiencies in

---

[53] GAO-14-555.

California's program. Specifically, reviews of well permitting files, including well location and information on aquifers surrounding the well, may have helped identify injections into nonexempt aquifers when compared to complete records on aquifer exemptions. However, according to EPA Region 9 officials, they have not conducted annual on-site evaluations of California's program. In 2011, regional officials requested a third-party audit of California's program, which was the first comprehensive review of California's program since primacy was granted in 1983. The audit found several program deficiencies, including inadequate inspection and enforcement practices and insufficient staff to adequately manage and implement the program, but Region 9 did not have complete information on approved aquifer exemptions in California and did not conduct a review of permitting files and aquifers in the area surrounding injection wells to identify wells that California had authorized to inject into nonexempt aquifers. According to EPA officials, in response to the recommendation from our June 2014 report for EPA to update its guidance on effective oversight, EPA headquarters and regional officials have held preliminary discussions to determine what oversight activities are necessary to ensure that state and EPA-managed programs are effective at protecting underground sources of drinking water, including on-site evaluations of state and EPA-managed programs.

Concerning why annual on-site reviews had not been consistently conducted, EPA headquarters and regional officials said that they have few resources to oversee state and EPA-managed programs, and regional officials told us that available resources are directed toward the class II programs they manage directly and not oversight of state programs. EPA headquarters officials we interviewed said that they have an effective oversight program and conduct necessary activities with the resources available. The same officials said they do not have the resources, including the workforce, necessary to consistently conduct the oversight activities to help assess whether state and EPA-managed programs are complying with applicable requirements.

According to a key workforce planning principle from our body of work on strategic human capital management, an agency should determine the critical skills and competencies that will be needed to achieve current and future programmatic results, particularly given factors that change the environment within which agencies work, such as budget constraints.[54] Our body of work on strategic human capital management indicates that each agency needs to ask if it has an explicit workforce planning strategy linked to the agency's strategic and program planning efforts to identify its current and future human capital needs, including the size of the workforce; its deployment across the organization; and the knowledge, skills, and abilities needed for the agency to pursue its shared vision.[55] In November 2015, EPA officials said that the agency had not conducted a comprehensive workforce analysis to identify the resources necessary, including human capital resources, to oversee state and EPA-managed programs, and that the agency had not requested additional resources for oversight. Without conducting such an analysis, EPA will not be able to identify the human capital and other resources it needs to carry out its oversight of state and EPA-managed programs and help ensure that they are effective at protecting underground sources of drinking water.

CONCLUSION

EPA established the UIC class II program in in the 1980s, with a vigorous role for the agency to oversee state and EPA-managed programs to prevent contamination of underground sources of drinking water. However, the findings in our June 2014 report, our findings on inspection and enforcement information and oversight activities in this report, and the recent decision that California's program was not complying with state and EPA requirements illustrate that EPA does not have the information, or consistently conduct the oversight activities, needed to assess state and

---

[54] GAO, *Human Capital: Key Principles for Effective Strategic Workforce Planning*, GAO-04-39 (Washington, D.C.: Dec. 11, 2003), and *Human Capital: A Self-Assessment Checklist for Agency Leaders*, GAO/OCG-00-14G (Washington, D.C.: September 2000).

[55] GAO/OCG-00-14G.

EPA-managed class II programs to help ensure that they protect underground sources of drinking water.

Specifically, the data EPA requires and collects from state and EPA-managed programs do not include well-specific information on inspections conducted by those programs needed to track each program's progress toward meeting its annual inspection goals, as called for in EPA's Program Reporting guidance. Until EPA requires and collects well-specific data on inspections from state and EPA-managed programs, including when wells were inspected, the types of inspections conducted at each well, and the results of those inspections, the agency does not have the well-specific information to assess whether the programs are meeting annual inspection goals to protect underground sources of drinking water.

To assess whether state and EPA-managed programs are effectively protecting underground sources of drinking water when permitting fluids to be injected into aquifers, EPA needs complete, updated information on approved aquifer exemptions. Yet EPA does not have a complete, up-to-date database on aquifer exemptions for all state and EPA-managed programs, or a way to keep the database containing information on aquifer exemptions updated. Until it has a complete aquifer exemption database and a way to update it, EPA does not have sufficient information on aquifer exemptions to oversee state and EPA-managed programs and assess whether programs are effectively protecting underground sources of drinking water.

Moreover, under the Safe Drinking Water Act, EPA must enforce state program requirements if they have not been enforced by the state in a timely and appropriate fashion. However, because of inconsistent interpretations of reporting guidance, state and EPA-managed programs report inconsistent and incomplete information on individual significant violations that have not been resolved, and therefore EPA regions and headquarters cannot know about, let alone take enforcement action against, operators committing significant violations. Until it clarifies guidance on what data should be reported on the 7520-4 form, EPA does not have reasonable assurance that state and EPA-managed programs report complete and consistent information on unresolved significant violations or that it has the information needed to assess whether it must take enforcement action, as required under the act, to protect underground sources of drinking water.

Finally, although EPA headquarters officials said they do not have the resources necessary to conduct the oversight activities needed to assess whether state and EPA-managed programs comply with applicable requirements, the agency has not conducted a workforce analysis to identify the resources, including human capital resources, the agency needs to oversee state and EPA-managed programs. Without conducting such an analysis, EPA will not be able to identify the human capital and other resources it needs to oversee state and EPA-managed programs and help ensure that they are effective at protecting underground sources of drinking water.

RECOMMENDATIONS FOR EXECUTIVE ACTION

To help ensure protection of underground drinking water from the injection of wastewater associated with domestic oil and gas production, we recommend that the Administrator of the Environmental Protection Agency take the following four actions:

- Require and collect well-specific data on inspections from state and EPA-managed programs, including when the wells were inspected, the types of inspections conducted, and the results of the inspections in order to track progress toward state and EPA-managed annual inspection goals.
- Complete the aquifer exemption database and establish a way to update it to provide EPA headquarters and regions with sufficient information on aquifer exemptions to oversee state and EPA-managed programs.
- Clarify guidance on what data should be reported on the 7520-4 form to help ensure that the data collected are complete and consistent across state and EPA-managed programs and to provide

the information EPA needs to assess whether it must take enforcement actions.

- Conduct a workforce analysis to identify the human capital and other resources EPA needs to carry out its oversight of state and EPA-managed programs.

AGENCY COMMENTS AND OUR EVALUATION

We provided the Administrator of EPA with a draft of this report for review and comment. In written comments provided by EPA (reproduced in app. VII), EPA generally agreed with our analysis and findings on the class II program and described planned actions, but disagreed with some findings and recommended actions, as discussed below. EPA also provided technical comments that we incorporated in the report, as appropriate. In addition, we provided the draft report to the six states whose programs we reviewed. Officials from these states—California, Colorado, North Dakota, Ohio, Oklahoma, and Texas—provided technical comments, which we incorporated as appropriate.

In response to our first recommendation that EPA require and collect well-specific data on inspections from state and EPA-managed programs to track progress toward state and EPA-managed annual inspection goals, EPA stated that the agency's goal is to obtain high quality data to understand program activities at the well-specific level, but that it didn't make sense to require the states to submit well-specific data now. EPA said that it is mindful of the need to think carefully about requiring information from states, and it will continue to work with its state partners to improve both the collection and the quality of the data currently required and to expand EPA's access to additional state data. Specifically, EPA stated that it had taken steps to address the gaps in the summary data collected on 7520 forms identified in GAO's June 2014 report, including developing standard operating procedures for submission and review of the data forms, and revising instructions to increase consistency in reporting the data to EPA. EPA said that it plans to continue to increase the inventory of well-specific data in the national UIC database including states that were working towards e-reporting status, and that EPA welcomes and encourages further participation. Further, EPA stated that it will continue to work with the Department of Energy and other stakeholders as they develop a database with well-specific state inventory data. We recognize EPA's efforts to improve the consistency and completeness of summary data collected on the 7520 forms, and to collect additional well-specific data through voluntary programs such as the national UIC database and the Department of Energy's database, but EPA has made little progress since 2007 collecting well-specific inspections data from state programs voluntarily. As we stated in the report, EPA needs access to well-specific inspections data from all programs to track the progress of state and EPA-managed programs towards meeting their inspection goals. If EPA believes that well-specific data is important, it should require that state and EPA-managed programs report well-specific data on inspections.

In response to our second recommendation that EPA complete the aquifer exemption database and establish a way to update it to provide EPA headquarters and regions with sufficient information on aquifer exemptions to oversee state and EPA-managed programs, EPA disagreed with our assessment that the agency is deficient in its duties to maintain aquifer exemption records, but is taking action to complete the database and to update it. Specifically, EPA stated that the draft report presents incomplete information as to which materials are held at the EPA headquarters and regional levels, and the roles and objectives that EPA headquarters and regions play regarding aquifer exemptions and the use of data. EPA said that our statement that the agency does not have sufficient information to oversee state and EPA-managed programs is incorrect because its regions have the most comprehensive and current data on aquifer exemptions as they conduct the final review of exemption requests and must approve all exemptions. According to EPA, it initiated the effort to collect data from the regional offices to better understand the number, locations, and nature and quality of aquifers exempted by the UIC program and expects to release a public data set by the end of 2016, which will include data current through 2015 with the exception of Region 9's data for the State of California. EPA stated that it anticipates adding Region 9's aquifer exemption data for California as the region works with the state to clarify the boundaries of the agency's historic approvals and takes action on the state's requests for new exemptions. Further, EPA said it plans to update the data set annually and that the regions will continue to hold the most current data. We commend EPA’s efforts to develop an up-to-date data set of aquifer exemptions and note that the updated information is important for overseeing whether the regions have current information on aquifer exemptions. As shown in the situation in Region 9 with California, at least one region did not have current or comprehensive information on aquifer exemptions. Further, EPA has been working since 2003 to compile comprehensive information on aquifer exemptions from regions, and, according to EPA officials, does not have a complete inventory of exemptions. In light of the situation in Region 9, until EPA has a complete aquifer exemption database and a way to update it, we continue to believe that it does not have sufficient information on aquifer exemptions to oversee state and EPA-managed programs and assess whether programs are protecting underground sources of drinking water.

In response to our third recommendation that EPA clarify guidance on what data should be reported on the 7520-4 form to help ensure that the data collected are complete and consistent across state and EPA-managed programs and to provide the information EPA needs to assess whether to take enforcement action, EPA agreed that the continued improvement in collection and consistency of data via the 7520-4 form would be valuable for more effective oversight. Specifically, EPA stated that the form is a tool for obtaining important information used in assessing enforcement activities and that providing guidance on the 7520-4 form could be valuable to improve the quality of information the agency receives. EPA also said that the 7520 standard operating procedures that it created in response to our June 2014 report reminds reviewers that wells with significant violations for two or more quarters should remain listed on the 7520-4 until the issue is resolved. In addition, EPA said that it will provide further materials to UIC data submitters to improve completeness and consistency of the data that programs report on the 7520-4 form within 6 months of this final report. As these standard operating procedures have not yet been finalized, we have not assessed them to determine whether they meet the intent of our recommendation.

In response to our fourth recommendation that EPA should conduct a workforce analysis to identify the resources it needs to conduct effective program oversight, EPA agreed that oversight is an important aspect of ensuring an effective UIC program, but stated that a workforce analysis was not necessary to better assess the resources needed to oversee the implementation of the UIC class II program. EPA stated that it is working with program managers to evaluate the effectiveness of EPA's oversight activities in response to our June 2014 report, and would expand the evaluation to include elements of inspection and enforcement activities if necessary. Upon completion of its evaluation, EPA said that it would look to improve the effectiveness of state and EPA oversight of the UIC programs, if needed. EPA may, for example, pilot a project to explore the potential to ensure program implementation by use of remote approaches, such as data collection, data analysis, targeting and priority ranking, and public transparency, as a viable option for increased oversight. While we recognize EPA's commitment to assess whether it should expand its evaluation of oversight activities to include inspections and enforcement, we still believe it is critical for EPA to identify the resources necessary, including human capital resources, to oversee state and EPA-managed programs and that without doing so, EPA may not have reasonable assurance that it can effectively collect information or conduct activities to ensure protection of underground sources of drinking water.

As agreed with your offices, unless you publicly announce the contents of this report earlier, we plan no further distribution until 30 days from the report date. At that time, we will send copies of this report to the appropriate congressional committees, the Administrator of the Environmental Protection Agency, and other interested parties. In addition, this report will be available at no charge on the GAO website at http://www.gao.gov.

If you or your staff members have any questions about this report, please contact me at (202) 512-3841 or gomezj@gao.gov. Contact points for our Offices of Congressional Relations and Public Affairs may be found on the last page of this report. Key contributors to this report are listed in appendix VIII.

J. Alfredo Gómez

Director,
Natural Resources and Environment

List of Requesters

The Honorable Barbara Boxer
Ranking Member
Committee on Environment and Public Works
United States Senate

The Honorable Sheldon Whitehouse
Ranking Member
Subcommittee on Fisheries, Water, and Wildlife
Committee on Environment and Public Works
United States Senate

The Honorable Edward Markey
Ranking Member
Subcommittee on Superfund, Waste Management, and
Regulatory Oversight
Committee on Environment and Public Works
United States Senate

The Honorable Peter DeFazio
Ranking Member
Committee on Transportation and Infrastructure
House of Representatives

The Honorable Raul Grijalva
Ranking Member
Committee on Natural Resources
House of Representatives

The Honorable Alan Lowenthal
Ranking Member
Subcommittee on Energy and Mineral Resources
Committee on Natural Resources
House of Representatives

The Honorable Diana DeGette
Ranking Member
Subcommittee on Oversight and Investigations
Committee on Energy and Commerce
House of Representatives

The Honorable Benjamin Cardin
United States Senate

APPENDIX I: LIST OF STATE AND EPA-MANAGED CLASS II PROGRAMS AND WELL INVENTORY

Table 3 provides a list of programs managed by the Environmental Protection Agency (EPA), state programs with safeguards deemed effective by EPA, state programs that have adopted minimum federal underground injection control requirements, and the number of class II wells in each state in 2014.

Table 3. Class II Program Management Authority and Well Inventory by State

State	EPA-managed programs	State programs with safeguards deemed effective by EPA	State programs that have adopted minimum federal requirements	Number of class II wells in FY2014
Alabama	NO	YES	NO	257
Alaska	NO	YES	NO	1,386
Arizona	YES	NO	NO	0
Arkansas	NO	YES	NO	1,118
California	NO	YES	NO	51,982
Colorado	NO	YES	NO	901
Connecticut	NO	NO	YES	0
Delaware	NO	NO	YES	0
Florida	YES	NO	NO	65
Georgia	NO	NO	YES	0
Hawaii	YES	NO	NO	0
Idaho	NO	NO	YES	0
Illinois	NO	YES	NO	8,167
Indiana	NO	YES	NO	1,242
Iowa	YES	NO	NO	7
Kansas	NO	YES	NO	16,947
Kentucky	YES	NO	NO	3,073
Louisiana	NO	YES	NO	3,755
Maine	NO	NO	YES	0
Maryland	NO	NO	YES	0
Massachusetts	NO	NO	YES	0
Michigan	YES	NO	NO	1,634
Minnesota	YES	NO	NO	0
Mississippi	NO	YES	NO	1,311
Missouri	NO	YES	NO	519
Montana	NO	YES	NO	1,206
Nebraska	NO	YES	NO	666
Nevada	NO	NO	YES	18
New Hampshire	NO	NO	YES	
New Jersey	NO	NO	YES	0
New MeYesico	NO	YES	NO	4,512
New York	YES	NO	NO	406
North Carolina	NO	NO	YES	0
North Dakota	NO	YES	NO	1,349
Ohio	NO	YES	NO	2,394
Oklahoma	NO	YES	NO	11,432
Oregon	NO	YES	NO	9

State	EPA-managed programs	State programs with safeguards deemed effective by EPA	State programs that have adopted minimum federal requirements	Number of class II wells in FY2014
Pennsylvania	YES	NO	NO	1,804
Rhode Island	NO	NO	YES	0
South Carolina	NO	NO	YES	0
South Dakota	NO	YES	NO	134
Tennessee	NO	NO	YES	29
TeYesas	NO	YES	NO	54,332
Utah	NO	YES	NO	640
Vermont	NO	NO	YES	0
Virginia	YES	NO	NO	15
Washington	NO	NO	YES	1
West Virginia	NO	YES	NO	402
Wisconsin	NO	NO	YES	0
Wyoming	NO	YES	NO	5,122

Legend: EPA = Environmental Protection Agency; FY = fiscal year. Source: EPA. | GAO-16-281.

Notes: This table excludes tribes, U.S. territories, and the District of Columbia. According to EPA, in some cases, states with no wells may have approval to manage their class II programs but may not have active class II programs.

APPENDIX II: STATUS OF CALIFORNIA'S UNDERGROUND INJECTION CONTROL CLASS II PROGRAM

According to a 2015 letter from California to Environmental Protection Agency (EPA) Region 9, California is the nation's third largest oil-producing state, producing 575,000 barrels per day, and the state's oil and gas industry earns $34 billion annually.[56] Injection wells have been used in the state for more than 50 years. According to a 2015 report, currently over 50,000 injection wells are operating in California, with about 75 percent of the state's production coming from enhanced oil recovery methods using underground injection wells.[57] California's class II underground injection control (UIC) program is managed by the Division of Oil, Gas, and

[56] California Division of Oil, Gas, and Geothermal Resources Letter to EPA Region 9 Re: Class II Oil and Gas Underground Injection Control (Feb. 6, 2015).

[57] California Division of Oil, Gas, and Geothermal Resources, *Underground Injection Control Program Report on Permitting and Program Assessment: Reporting Period of Calendar Years 2011-2014* (Sacramento, CA: October 2015).

Geothermal Resources (Division) and is divided across the Division's six district offices. The majority of class II underground injection activity occurs in District 1 (Cypress) and District 4 (Bakersfield).

California Program Determined to Be Noncompliant

In July 2014, EPA Region 9 determined that the UIC class II program managed by the Division did not comply with state and EPA requirements. In a series of letters from July 2014 through July 2015, EPA Region 9 and the Division reached agreement on a plan to improve California's UIC class II program. Below is a summary of the deficiencies identified in California's UIC class II program and the plans California and EPA Region 9 agreed on to resolve these deficiencies, including actions taken by EPA and California before and after the determination of noncompliance in 2014.

- In 2011, EPA requested a third-party audit of the state's UIC class II program.[58] The audit made recommendations to improve California's class II program, including recommendations regarding the program's definition of underground sources of drinking water, area of review calculations, well construction practices, inspection and enforcement practices, and staff qualifications.
- In November 2012, the Division developed an action plan to address each of the recommendations from EPA's audit. To address a number of recommendations necessitating regulatory updates, the Division committed to update its class II program regulations beginning in 2013.[59]

[58] Horsely Witten Group for EPA Region 9, *California Class II Underground Injection Control Program Review* (June 2011).

[59] California Division of Oil, Gas, and Geothermal Resources, *Underground Injection Control Action Plan: Response to the U.S. EPA June 2011 Review of California's UIC Program* (Sacramento, CA: November 2012).

- In response to an EPA inquiry initiated in 2012, California reviewed program records to ensure that injection wells the state authorized aligned with EPA-approved aquifer exemptions.[60] In doing so, in 2014 the Division discovered that it authorized operators to inject class II wastewater into 11 nonexempt aquifers in the vicinity of water supply wells, and EPA determined that the program was not in compliance with state and EPA requirements.
- In October 2015, the Division issued the first report from its Monitoring and Compliance Unit, which was created in 2011.[61] The report identified a number of program deficiencies, including insufficient staffing to address increasing regulatory workload and significant remedial programmatic work; poor recordkeeping on mostly paper forms and a lack of modern data tools and systems; outdated regulations that in some cases do not address the modern oil and gas extraction environment; inconsistent and understaffed program leadership; insufficient breadth and depth of technical talent; insufficient coordination among district and state offices; and lack of consistent, regular, high-quality technical training.

Division officials also identified deficiencies with the enforcement of class II requirements. Division officials said that the state office receives violation information from districts and is responsible for pursuing enforcement actions against operators and collecting penalties assessed by the Division. However, according to Division officials, California has historically had difficulties enforcing regulations for both production and class II wells in the state. In particular, the Division identified many examples of enforcement actions that were not pursued and wells that were

---

[60] If certain conditions are met, aquifers can be exempted from protection under the Safe Drinking Water Act. Well operators may request an exemption for a particular aquifer, and if EPA approves, operators may inject fluids into the aquifer.

[61] In 2011, the Division created the Monitoring and Compliance Unit to evaluate program compliance with state and EPA requirements. The Monitoring and Compliance Unit was tasked with evaluating and reporting on the strengths and challenges of the state's program in meeting the statutory and regulatory standards on which the program is based, including state statutes and regulations and California's memorandum of agreement with EPA detailing how the state would manage its program to comply with the Safe Drinking Water Act.

not being returned to compliance in a timely manner. For example, in 2010, the Division hired a contractor to review its accounts receivable to identify outstanding penalties that the Division had not collected. According to Division officials, there were over $5 million in unpaid penalties that the Division had assessed but did not collect. In September 2015, according to Division officials, the Division hired a deputy supervisor to start tracking enforcement of state requirements and to lead the development of new business processes to improve violation tracking and enforcement.

Division's Plan for Program Improvements

Since July 2014, the Division, California's State Water Resources Control Board (Board), and EPA have been working together to systematically address a number of important deficiencies in the UIC program, including permitting injection into nonexempt aquifers.[62] In letters between California (the Division and the Board) and EPA, the three-agency group agreed to a plan for the Division to shut down wells permitted to inject into nonexempt aquifers and improve and modernize its UIC practices.[63] Specifically, the plan consists of four major components to be completed concurrently:

- **New regulations and program revisions.** The Division determined that many state regulations that govern underground injection control are obsolete, deficient, or unable to address current industry practice. According to agency documents, the Division plans to undertake a series of rulemakings to improve California's regulatory framework to address these issues, including isolation of injected fluids, quality of water to be

[62] The Board consults with the Division on injection well permits.

[63] California Division of Oil, Gas, and Geothermal Resources, *Underground Injection Control Program Report on Permitting and Program Assessment: Reporting Period of Calendar Years 2011-2014*.

protected, well construction practices, cyclic steam operations,[64] project review, and idle well standards and testing. In July 2015, the Division stated that it planned to update its class II regulations in two phases, with the first phase starting with the informal circulation of draft regulations in the fall of 2015 and the second phase beginning in 2016.[65]

- **Well review and aquifer exemptions.** The Division and the Board have been systematically reviewing injection wells that may have been permitted to inject into nonexempt aquifers. The Division has proposed a schedule for reviewing and ceasing injection into these aquifers. As of October 2015, the Division shut down 23 wells injecting wastewater into underground sources of drinking water that may have posed an immediate risk to waters of beneficial use. Over the next 2 years, through 2017, according to agency documents, the Division will review additional injection wells to determine whether they should be shut down or continue operating.[66] The Division is collecting information from operators interested in pursuing exemptions and will review each exemption application to determine whether exemption criteria have been sufficiently met. If the Division approves the aquifer exemption, it will forward it to EPA for review and approval or disapproval. EPA has final authority to declare an aquifer exempt. The Division has issued regulations to ensure that injection activity ends by specified deadlines unless aquifer exemptions are approved.
- **Project-by-project review of injection project approvals.** The Division plans to conduct individual project reviews designed to find missing data, identify UIC compliance issues, and compare

---

[64] Cyclic steam wells inject steam into hydrocarbon producing formations to enhance recovery of oil.

[65] California Division of Oil, Gas, and Geothermal Resources, *Underground Injection Control Program Report on Permitting and Program Assessment: Reporting Period of Calendar Years 2011-2014.*

[66] California Division of Oil, Gas, and Geothermal Resources, *Underground Injection Control Program Report on Permitting and Program Assessment: Reporting Period of Calendar Years 2011-2014.*

existing project approvals with current conditions in the field.[67] Operators will be required to provide missing data, and the Division will reevaluate the project based on all relevant regulations, mandates, and policies, including demonstration of zonal isolation of injected fluids. Projects will be reapproved, modified, or canceled as appropriate. The Division plans to conduct separate reviews in each Division district and plans to complete the review by October 2018.

- **Development of a modern well and data management system.** The Division is updating its data management systems for production and injection wells to improve regulatory compliance and effectiveness, transparency, and support of all stakeholders. Finishing every component of the UIC improvement plan submitted to EPA could take 3 to 4 years. However, according to state documents, as each piece is completed, improvements in the Division's mission performance will follow. According to state documents, changes will be supported by the development of training programs to support the process of internal review and adjustments for continuously improving the Division's execution of its responsibilities.

APPENDIX III: OBJECTIVES, SCOPE, AND METHODOLOGY

This report examines the Environmental Protection Agency's (EPA) Underground Injection Control (UIC) class II program to determine the extent to which EPA has collected the inspection and enforcement information needed, and conducted the oversight activities necessary, to assess that state and EPA-managed programs are protecting underground

[67] A project under the Division's class II program consists of many wells, sometimes as many as 200 wells, in an injection production system. A project includes both injection and production wells. The project proposal includes evaluation of the geology of the area to be subject to injection and production operations. It also must include review of the construction of neighboring wells and the ability of the geologic structures to contain injection fluid within the intended injection zone.

sources of drinking water. To perform this work, we reviewed and analyzed the Safe Drinking Water Act, and EPA regulations and guidance applicable to the UIC class II program. We also interviewed EPA UIC program officials in the eight regional offices with class II wells. To understand the class II program at the state level, we interviewed state officials and reviewed state program documentation for the same sample of states from our June 2014 report on the UIC program.[68] Specifically, we selected a nongeneralizable sample of eight states with class II programs. Two of these states are managed by EPA regions—Kentucky and Pennsylvania—and the remaining six—California, Colorado, North Dakota, Ohio, Oklahoma, and Texas—are managed under provisions of the act that allow them to have primary responsibility to manage the program in their states. We selected these states from the six shale oil and gas regions defined by the Energy Information Administration. For each of the six shale regions, we selected at least one state that had among the highest number of class II injection wells.

In July 2014, after we issued our June 2014 report and before we started the work on this review, EPA determined that one of the programs in the eight states we reviewed, California's class II program, was not in compliance with state or EPA requirements. EPA Region 9 officials and California's UIC program officials have since agreed to a plan to improve the California program over the next several years. We interviewed EPA headquarters, EPA Region 9, and California officials regarding the deficiencies in California's program, the agreed-upon improvement plan, and EPA oversight of California's progress. A summary of the deficiencies found by EPA and California, and California's plans to improve its program, can be found in appendix II. Because of the deficiencies in California's program, we chose not to include California in our detailed analysis of inspection and enforcement information from the states. Thus, the results of our review of inspection and enforcement reflect the seven states remaining in our sample. Because the sample is a nongeneralizable

---

[68] GAO, *Drinking Water: EPA Program to Protect Underground Sources from Injection of Fluids Associated With Oil and Gas Production Needs Improvement*, GAO-14-555 (Washington, D.C.: June 27, 2014).

sample, our results cannot be generalized to other states but do provide detailed examples of EPA's and states' management of class II programs.

To analyze whether EPA collects the information it needs to assess whether state and EPA-managed programs are protecting underground sources of drinking water, particularly inspection and enforcement information, we first reviewed EPA regulations and guidance on UIC inspections and enforcement to determine what information EPA needs to assess the programs and their ability to protect underground sources of drinking water. EPA's 1987 guidance document Underground Injection Control Program Compliance Strategy for Primacy and Direct Implementation Jurisdictions (Strategy) establishes minimum goals for inspections of class II wells.[69] We obtained and summarized inspections data collected by EPA from each program we reviewed for fiscal year 2013, the most current year of data available at the beginning of this review. The state and EPA-managed programs are directed to report these data to EPA quarterly on the 7520-3 form. To assess the reliability of these data, we interviewed EPA and state officials about their processes for managing the data collected on the 7520-3 forms and tested the data for completeness. We found that the data were not comparable across states but were sufficiently reliable for reporting on a state-by-state basis. To understand EPA's use of the data to assess state and EPA-managed programs, we interviewed officials from EPA headquarters about their use of the information to oversee EPA-managed programs and from EPA regions about their oversight of inspections conducted by state programs. We also interviewed selected state program officials about how they manage class II inspections, and we requested information on annual inspection goals and inspection strategies. Similarly, we interviewed regional office staff responsible for managing the programs in Kentucky and Pennsylvania about their management of the class II programs in these states, including any inspection goals and strategies they have.

To analyze whether EPA has the enforcement information to assess whether state and EPA-managed programs are protecting underground

---

[69] EPA, Office of Drinking Water, *Underground Injection Control Program Compliance Strategy for Primacy and Direct Implementation Jurisdictions* (Washington, D.C.: Mar. 31, 1987).

sources of drinking water, we reviewed EPA's Strategy, which also establishes enforcement expectations for both state and EPA-managed programs. In particular, the Strategy identifies the need for state and EPA-managed programs to conduct timely and appropriate enforcement actions. Specifically, state and EPA-managed programs are expected to resolve significant violations within 90 days of discovery or take a formal enforcement action against the well operator. According to the Strategy, a formal enforcement action, among other things, is legally enforceable, explicitly requires the well owner to take corrective action, and specifies a timetable for completion. Under the act, EPA is to intervene and take enforcement action once it is notified that a violation has occurred and that the state has not taken appropriate action after 30 days. Similarly, EPA regions should take timely and appropriate enforcement actions in states with EPA-managed programs. According to EPA's 1987 Reporting Requirements—Underground Injection Control Program Guidance (Program Reporting),[70] EPA uses the 7520-4 forms to evaluate the timeliness and appropriateness of a state or EPA-managed program's enforcement response; EPA regions receive 7520-4 forms from their state programs, and EPA headquarters collects information on the 7520-4 forms from programs managed by EPA regions.

We then assessed a sample of violations, using EPA's definition of timely and appropriate resolution from its Strategy and Program Reporting guidance, to determine if EPA receives information on individual significant violations that may have the potential to threaten underground sources of drinking water. We selected a nongeneralizable sample of 134 notices of violation, issued from 2008 through 2013 (the most recent years of data available when we began our audit work), from the seven state and EPA-programs we reviewed and compared the data to enforcement data provided to EPA on the 7520-4 forms.[71] We selected a nongeneralizable

[70] EPA, Office of Drinking Water, *FY 1987 Reporting Requirements—Underground Injection Control Program Guidance (UICP) Guidance #53* (Washington, D.C.: December 1986).

[71] For the purposes of this report, we refer to all written notifications to operators that they are in violation of state or EPA requirements as notices of violation. During our review of seven states, we noted that state or EPA-managed programs may initiate a single enforcement case

sample of at least six notices of violation in each of the seven states in our sample based on the significance of the violation; the type of enforcement action taken; and the number of days between when the operator was notified and when the violation was resolved, termed returning to compliance with applicable requirements. We also obtained the 7520-4 forms for fiscal years 2008 through 2013 to identify what violations had been reported on these forms. We analyzed the number of days that each significant violation in our sample had been open and compared this to the number of days (90) established by EPA as timely.[72] We then analyzed each violation to determine if a formal action had been taken. We identified 93 violations that were open for more than 90 days and compared these to the information reported on the 7520-4 form by the appropriate state. We then interviewed EPA, regional, and state officials to determine how they reported the information on the 7520-4 form.

Because our sample of violations is nongeneralizable, our results cannot be generalized to other states and violations; however, they do provide detailed information on the violations that should have been reported by state and EPA-managed programs. To assess the reliability of the violation and enforcement information we obtained, we interviewed EPA headquarters officials about their processes for collecting and managing the information and tested the information for completeness by looking for missing information. We determined that the information from EPA's reporting forms was reliable for purposes of reporting individual state results.

To analyze the activities EPA conducts to assess whether state and EPA-managed programs protect underground sources of drinking water, we reviewed several EPA guidance documents that describe activities EPA is to take to oversee state and EPA-managed programs. EPA's 1983 guidance document, Interim Guidance for Overview of the Underground Injection Control Program, states that EPA is supposed to conduct annual

---

against an operator for multiple violations, so a single notice of violation may cover more than one violation.

[72] Significant violations, or significant noncompliance, are violations that in general, pose a threat to underground sources of drinking water.

on-site evaluations of state and EPA-managed programs.[73] EPA's UIC regulations describe activities that EPA is supposed to conduct to ensure that it can enforce state program requirements, if necessary. EPA's 1984 guidance, Guidance for Review and Approval of State Underground Injection Control Programs and Revisions to Approved State Programs, describes the activities that EPA is to conduct to review changes to state program requirements.[74] We reviewed the extent to which EPA conducted the first two activities in our June 2014 report on the UIC program.[75] We met with EPA headquarters officials to discuss our findings from that report and EPA's efforts to implement our recommendations.

To analyze the extent to which EPA has carried out activities to review and approve aquifer exemptions for state and EPA-managed programs, we reviewed EPA guidance documents on aquifer exemptions. We then interviewed EPA headquarters officials about EPA's progress developing and maintaining a database on aquifer exemptions. To analyze the extent to which EPA applied best practices for workforce planning and strategic human capital management to the management of the UIC program, we reviewed GAO reports specifying best practices for strategic human capital management. We then interviewed EPA's headquarters officials about EPA's efforts to apply those best practices to the UIC program.

We conducted this performance audit from October 2014 to February 2016 in accordance with generally accepted government auditing standards. Those standards require that we plan and perform the audit to obtain sufficient, appropriate evidence to provide a reasonable basis for our findings and conclusions based on our audit objectives. We believe that the evidence obtained provides a reasonable basis for our findings and conclusions based on our audit objectives.

---

[73] EPA, Office of Drinking Water, *Interim Guidance for Overview of the Underground Injection Control (UIC) Program, Groundwater Program Guidance #30* (Washington, D.C.: June 1983).

[74] EPA, Office of Drinking Water, *Guidance for Review and Approval of State Underground Injection Control (UIC) Programs and Revisions to Approved State Programs GWPB Guidance #34* (Washington, D.C.: July 1984).

[75] GAO-14-555.

Appendix IV: Information on Inspections Conducted by Selected State and EPA-Managed Programs

The Safe Drinking Water Act requires states to include inspection requirements in their programs. In a 1987 document titled Underground Injection Control Program Compliance Strategy for Primacy and Direct Implementation Jurisdictions (Strategy), the Environmental Protection Agency (EPA) provides guidance on the types of inspections and the frequency at which they should be conducted.[76] The types of inspections include routine inspections, well construction inspections, mechanical integrity tests witnessed, emergency and complaint response, enforcement follow-up, and plugging and abandonment verification.

According to the Strategy, the goal of an inspection program is to determine that a well is in compliance with applicable requirements and to detect any violations of those requirements. It directs programs to adopt minimum priority standards for each type of inspection and gives the programs discretion to consider additional priorities, such as environmental risks, population risks, and well construction, when determining which wells to inspect. The Strategy ranks inspection types, including those for class II wells, by priority, as shown in *Table 4*.

The seven state and EPA-managed programs we reviewed establish goals for each of the inspection types identified in the Strategy based on program priorities and available inspection resources.[77] *Table 5* shows state and EPA-managed program inspection goals by inspection type. Some states have goals to inspect all of their wells monthly or quarterly. For example, North Dakota program officials told us that their goal is to conduct routine inspections at all class II injection wells monthly, and Ohio

---

[76] EPA, Office of Drinking Water, *Underground Injection Control Program Compliance Strategy for Primacy and Direct Implementation Jurisdictions* (Washington, D.C.: Mar. 31, 1987).

[77] We included seven states in our analysis of inspection and enforcement data because we dropped one of the state programs from our initial sample of eight state and EPA-managed programs. We chose not to include California in our detailed analysis of inspection and enforcement information from the states because EPA determined in July 2014 that California's program was not in compliance with state and EPA requirements.

has a goal of inspecting 100 percent of its wells quarterly, according to program officials. Other programs we reviewed do not set specific annual goals for individual types of well inspections. For example, according to EPA Region 4 officials, EPA Region 4 has a goal of conducting routine inspections of all of the class II wells in Kentucky at least once every 5 years and does not set inspection goals for observing well plugging and well construction. Some state and EPA program officials told us that when states do not have the resources to inspect all wells annually, type and frequency of inspections are prioritized based on risk factors such as the operator's history of compliance with state or EPA requirements or danger to the general public. For example, Oklahoma officials told us that inspectors will inspect an operator more frequently if the inspector determines the well operator is violating state requirements and will also prioritize inspections in areas of the state where there is a history of illegal disposal activity. Similarly, EPA Region 3 officials told us that they do not set annual inspection goals by inspection type, but prioritize inspections based on factors such as danger to the general public, emergency response, and the availability of inspection staff.

Generally, according to state officials, state inspection staff in the five state programs we reviewed are responsible for inspecting both production and class II injection wells in the state. According to state officials, staff may inspect only class II or production wells or may inspect both. For example, of the 47 staff members conducting inspections in Ohio, 4 conduct inspections on class II wells full-time and 8 to 10 split responsibilities between production and class II wells. Similarly, inspection staff in EPA regions we reviewed are responsible for inspecting all classes of injection wells managed by the region.[78] For example, according to EPA

---

[78] EPA regulates six classes of underground injection wells. Class II wells are used to inject brines and other fluids associated with oil and gas production, and hydrocarbons for storage, and are the focus of this report. Class I wells are used to inject hazardous wastes, industrial nonhazardous liquids, or municipal wastewater beneath the lowermost underground drinking water sources. Class III wells are used to inject fluids associated with solution mining of minerals beneath the lowermost underground drinking water source. Class IV wells are used to inject hazardous or radioactive wastes into or above underground drinking water sources (these wells are banned unless authorized under a federal or state groundwater remediation project). Class V wells, in general, are used to inject nonhazardous fluids into or above

Region 4 officials, Region 4 has approximately 3 program staff members and 2 contractors to conduct inspections of all classes of injection wells in the region.

Table 4. Types of Class II Well Inspections and EPA-Recommended Annual Inspection Goals by Percentage of Wells Inspected or Frequency of Inspections

Inspection type	Priority (1 = high; 4 = low)	Description	Recommended annual inspection goals by percentage or frequency
Emergency and complaint response	1	Response to an emergency situation that constitutes imminent hazard or to citizen complaints where the program has reason to believe there is potential endangerment to underground sources of drinking water.	100%
Mechanical integrity tests witnessed	2	Witness tests to ensure that there are no significant leaks from well andnosignificant fluid movement into underground sources of drinking water.	25%
Enforcement inspection	2	Inspections to support enforcement activity, including follow-up to enforcement actions.	100%
Inspection type	Priority (1 = high; 4 = low)	Description	Recommended annual inspection goals by percentage or frequency
Plugging and abandonment verification	3	Verification that the owner/operator has complied with all program requirements associated with plugging and abandoning a well.	25%
Well construction inspection	3	Verification of adequate well construction and engineering prior to start-up to ensure that new and reconditioned wells are in compliance with program requirements.	As resources allow
Routine inspection	4	Inspection performed routinely or in response to a complaint to ensure that the well operator is in full compliance with program requirements.	At least once every 5 years

Legend: EPA = Environmental Protection Agency.

Source: GAO analysis of EPA guidance. | GAO-16-281.

---

underground drinking water sources and are typically shallow, on-site disposal systems. Class VI wells are used to inject carbon dioxide for long-term storage.

Table 5. Reported Annual Inspection Goals for Selected Programs by Class II Well Inspection Type by Percentage of Total Inspected or Number of Inspections Conducted

State	Emergency and complaint response	Mechanical integrity tests witnessed	Enforcement	Well pluggings witnessed	Well constructions witnessed	Routine inspections
EPA Strategy	100%	25%	100%	25%	As resources allow	At least once every 5 years
Colorado	100%	100%	100%	100%	100%	100%
Kentucky	100%	100%	100%	None	None	At least once every 5 years
North Dakota	100%	100%	100%	100%	100%	100%
Ohio	100%	100%	100%	100%	Highest possible percentage	100%
Oklahoma	100%	90%	100%	As resources allow	As resources allow	At least 10,000
Pennsylvania	None	None	None	None	None	None
Texas	100%	8,000	100%	65%	1,896	None[a]

Legend: EPA = Environmental Protection Agency; Strategy = Underground Injection Control Program Compliance Strategy for Primacy and Direct Implementation Jurisdictions. Source: GAO analysis of EPA and selected state data. | GAO-16-281

Notes: We chose not to include California in our detailed analysis of inspection and enforcement information from the states because EPA determined in July 2014 that California's program was not in compliance with state and EPA requirements. The results of our review of inspection and enforcement information reflect the seven remaining state and EPA-managed programs we reviewed. Kentucky and Pennsylvania are EPA-managed programs.

[a]Texas has an annual goal of 118,800 routine well inspections, which includes but is not limited to class II wells.

State agencies and EPA regional offices responsible for managing or overseeing programs in the seven states we selected vary in the inventory of wells they manage and the staffing resources dedicated to inspect those wells. For example, according to North Dakota officials, North Dakota has 35 staff members to inspect 14,158 production and class II injection wells in the state. According to Oklahoma officials, Oklahoma has 62 staff members to inspect the state's almost 190,000 production and class II wells. EPA regions managing programs in states we selected have comparatively fewer inspection staff to inspect the injection wells they manage. For example, according to EPA officials, Region 3 has 1 full-time inspector and 3 part time inspectors to inspect the almost 29,000 injection wells it manages region-wide, including over 1,800 class II wells in Pennsylvania.

APPENDIX V: INFORMATION ON ENFORCEMENT ACTIONS TAKEN BY SELECTED STATE AND EPA-MANAGED PROGRAMS

This appendix contains information on the enforcement process used by selected state programs and programs managed by the Environmental Protection Agency (EPA), the enforcement tools available to programs, and our analysis of a sample of enforcement cases that we reviewed. We selected a nongeneralizable sample of at least six enforcement cases in each of the seven states in our sample,[79] based on the significance of the violation, the type of enforcement action taken, and the number of days between when the operator was notified and when the violation was resolved— defined as returning to compliance—with applicable requirements.

Enforcement Process

According to EPA's Underground Injection Control Program Compliance Strategy for Primacy and Direct Implementation Jurisdictions (Strategy), state and EPA-managed programs should escalate their

---

[79] We included seven states in our analysis of inspection and enforcement data because we dropped one of the state programs from our initial sample of eight state and EPA-managed programs. We dropped California from our analysis of inspection and enforcement data because EPA determined in July 2014 that California's program was not in compliance with state and EPA requirements.

enforcement response if compliance is not achieved in a timely manner.[80] The enforcement action taken can depend on a number of factors, including the severity of the violation and its potential to contaminate underground sources of drinking water. Our analysis of enforcement actions taken by select state and EPA-managed programs found that the programs have generally similar enforcement processes to respond to class II Underground Injection Control (UIC) violations.

A violation can be discovered a number of ways, including through an inspection, administrative review of a well file, or reports by citizens or others.[81] According to EPA and state officials, the enforcement process generally begins when program officials notify a well operator that the well is in violation of applicable requirements. Six of the seven programs we reviewed generally issue a written notice of the violation to the well operator, but North Dakota program officials told us that they instead give a verbal notification and then, depending on the severity of the violation, will allow a 30-day grace period before initiating a formal enforcement action. For the state and EPA-managed programs we reviewed, notices of violation can include one or more violations in a single notice. If the operator does not take action to resolve a significant violation—that is, return the well to compliance with all state and federal regulations—in a timely manner, EPA's 1987 Strategy directs state and EPA-managed programs to take formal enforcement actions to ensure that compliance is achieved.[82] Formal enforcement actions can include the following:

Administrative Orders

Administrative orders are legally enforceable orders, the terms of which can either be dictated by the program or negotiated with the well operator in violation (which may be referred to as a consent order or

---

[80] EPA, Office of Drinking Water, *Underground Injection Control Program Compliance Strategy for Primacy and Direct Implementation Jurisdictions* (Washington, D.C.: Mar. 31, 1987).

[81] A violation can be reported to the state or EPA-managed program by a citizen or citizen group. Also, a well operator may self-report a violation to its respective class II program.

[82] EPA, Office of Drinking Water, *Underground Injection Control Program Compliance Strategy for Primacy and Direct Implementation Jurisdictions*.

consent agreement).[83] Administrative orders may enjoin the well operator from taking certain actions, may require the well operator to take corrective action, and may impose monetary penalties.

Civil Judicial Action

Civil judicial actions are lawsuits filed against an operator that has failed to comply with, for example, statutory or regulatory requirements or an administrative order. Civil actions are generally taken when administrative enforcement actions have been unsuccessful in achieving compliance and resolving the violation, according to EPA officials.

Criminal Judicial Action

A program may also refer a case to the criminal justice system if an action is willfully committed. A criminal court conviction can result in fines or imprisonment.

Enforcement Tools Available to State and EPA-Managed Programs

State and EPA-managed programs have various tools available to facilitate a return to compliance with applicable requirements and deter future violations. According to EPA officials, state and EPA-managed programs can vary in their approaches to enforcing UIC program requirements as long as the programs are effective at protecting underground sources of drinking water. Six of the seven state and EPA-managed programs in our review have authority to assess monetary penalties. *Table 6* details the types of administrative, civil, and criminal monetary penalty authority available at the state and federal levels for the selected states in our review.

---

[83] Consent orders offer terms that are generally agreed upon by the program and the well operator.

Table 6. Penalties for Violations of Selected State and EPA-Managed Program Requirements

State	Administrative penalty	Civil penalty	Criminal penalty/imprisonment
Colorado	Maximum penalty of $15,000 per violation per day.	None specified.	Maximum penalty of $5,000, 6 months' imprisonment, or both.
Kentucky	Up to $10,000 fine per violation per day up to a maximum of $125,000.	Maximum penalty of $25,000 fine per violation per day.	Maximum penalty of $25,000 fine per violation per day, 3 years' imprisonment, or both. Additional penalties can be imposed under Title 18 of the United States Code.
North Dakota	Maximum penalty of $12,500 per violation per day.	None specified.	Maximum penalty of $10,000, 5 years' imprisonment, or both.
Ohio	None identified.	$2,500 to $20,000 per violation per day for disposal wells; up to $10,000 per violation per day for enhanced recovery wells.	$10,000, 6 months' imprisonment, or both for the first disposal well offense; $20,000, 2years' imprisonment, or both for subsequent offenses.
Oklahoma	Maximum penalty of $5,000 per violation per day.	None specified.	Maximum penalty of $5,000, 5 years' imprisonment, or both.
Pennsylvania	Up to $10,000 fine per violation per day up to a maximum of $125,000.	Maximum penalty of $25,000 fine per violation per day.	Maximum penalty of $25,000 fine per violation per day, up to 3 years' imprisonment, or both. Additional penalties can be imposed under Title 18 of the United States Code.
Texas	Maximum penalty of $10,000 per violation per day.	Maximum penalty of $5,000 per act of noncompliance per day.	Maximum penalty of $10,000 per violation per day. Falsification of applications, reports, or documents or tampering with gauges may also carry a $10,000 penalty, imprisonment of 2 to 5 years, or both.

Legend: EPA = Environmental Protection Agency.

Source: GAO analysis of selected state and EPA-managed program requirements. | GAO-16-281.

Notes: We chose not to include California in our detailed analysis of inspection and enforcement information from the states because EPA determined in July 2014 that California's program was not in compliance with state and EPA requirements. The results of our review of inspection and enforcement information reflect the seven remaining state and EPA-managed programs we reviewed. Kentucky and Pennsylvania are EPA-managed programs.

While six of the seven programs we reviewed have the legal authority to assess monetary penalties, some do not regularly use these authorities for various reasons. For example, North Dakota's program has administrative authority to assess a monetary penalty, but the state prefers to employ a more cooperative approach to get operators to bring wells back into compliance, according to North Dakota officials. According to Colorado officials, Colorado's program has also historically employed a cooperative approach, but the state recently revised its regulations to require a more prescriptive approach to enforcement. Ohio program officials told us that they do not have an administrative process for assessing a monetary penalty, and the penalty must instead be pursued through the civil judicial process. In turn, Ohio officials told us that they consider the advantages and disadvantages of resolving a violation through negotiated consent agreement before referring the case to the state's attorney general to pursue civil penalties.

Other tools available to selected state and EPA-managed programs to enforce program requirements may include the following:

Well Shut-In

Some programs we reviewed may temporarily close down a well until a violation is resolved. For example, Oklahoma officials can shut-in a well if an operator is out of compliance with its financial assurance requirements.

Pipeline Severance

A program may also have the authority to sever an operator's access to oil and gas pipelines. For example, if an operator uses a well that has been shut-in for violations, Texas may take the further step of refusing to renew certain documents the operator needs to do business in the state. If the disposal well operator also has production wells in the state, this would prevent the operator from producing oil and gas. According to EPA officials, this can be an effective enforcement tool given that a company's income is generated on the production side.

Permit Revocation or Temporary Suspension

A program may have the authority to revoke or temporarily suspend existing UIC permits, thereby making it illegal for an operator to continue injecting into a well or group of wells covered under the permit. For example, North Dakota may revoke permits after notice and hearing if the well operator fails to comply with the terms and conditions of its permit or any applicable rule or law, and the state may suspend permits for good cause.

Moratorium on New or Renewed Permits

A program may be able to refuse to issue new permits to an operator with a history of noncompliance. For example, Oklahoma program officials can seek an order denying a permit to an operator with an unsatisfactory compliance history.

Bond Forfeiture

A program may require well operators to post a bond to ensure compliance with requirements applicable to the well. If an operator fails to comply with these terms, a state may be able to seize the bond to cover the costs of returning the well to compliance. For example, if Ohio officials find that an operator has failed to comply with, among other things, certain orders, regulations, or its permit, it may declare the operator's bond to be forfeit.

Timely and Appropriate Enforcement of State and EPA Requirements

Under the Safe Drinking Water Act, EPA must enforce state requirements if violations have not been enforced by states in a timely and appropriate manner. EPA's Strategy sets forth standards for timely and appropriate enforcement action in response to significant violations. Specifically, state and EPA-managed programs are expected to resolve significant violations within 90 days of discovering the violation or take a formal enforcement action against the well operator. According to the Strategy, a formal enforcement action, among other things, is legally enforceable, explicitly requires the well owner to take corrective action, and specifies a timetable for completion. When EPA becomes aware that an operator is violating a state program requirement, a provision in the act requires EPA to notify the state and, if the state does not take appropriate action within 30 days, to intervene by issuing an administrative order or commencing a civil action. Similarly, EPA regions should take timely and appropriate enforcement actions in states with EPA-managed programs.

Table 7. Summary of Types of Violations in 134 Notices of Violation from Selected State and EPA-Managed Programs Reviewed for Fiscal Years 2008 through 2013

State	Unauthorized injection	Mechanical integrity	Injection over pressure	Failure to plug	Violation of formal order	Pollution	Reporting	State
Colorado	1	9	1	1	0	0		3
Kentucky	18	11	1	1	0	0		2
North Dakota	1	19	1	4	0	0		1
Ohio	0	16	1	2	0	6		0
Oklahoma	17	21	1	0	1	1		16
Pennsylvania	4	2	0	0	0	0		0
Texas	3	17	0	16	0	0		0
Total	44	95	5	24	1	7		22

Legend: EPA = Environmental Protection Agency.

Source: GAO analysis of EPA and selected state data. | GAO-16-281.

Notes: Although California was one of the eight states we selected, we chose not to include California in our detailed analysis of inspection and enforcement information from the states because EPA determined in July 2014 that California's program was not in compliance with state and EPA program requirements. The results of our review of inspection and enforcement information reflect the seven remaining state and EPA-managed programs we reviewed. Some state or EPA-managed programs may initiate a single enforcement case against an operator for multiple violations, so a single enforcement case may cover more than one type of violation, which may result in more violations than enforcement cases from each program. Kentucky and Pennsylvania are EPA-managed programs.

State and EPA-managed programs are required to submit periodic reports to EPA headquarters with information on enforcement actions taken against well operators. One of the required reports is to provide quarterly information on individual significant violations by well operators that have not been resolved and that may have the potential to threaten underground sources of drinking water. EPA uses the 7520-4 form to collect this information.[84] According to EPA's 1987 Reporting Requirements—Underground Injection Control Program Guidance (Program Reporting),[85] EPA uses the 7520-4 forms to evaluate the timeliness and appropriateness of a state or EPA-managed program's enforcement response.

Our analysis of a sample of significant violations from selected state and EPA-managed programs found that a subset of significant violations that should have been reported on the 7520-4 forms were not reported, and that the forms contained incomplete and inconsistent information. Specifically, we sampled 134 notices of violation from selected state and EPA-managed programs, of which 93 included significant violations (see app. VI for a list of the enforcement cases we reviewed in the seven state and EPA-managed programs).[86] *Table 7* shows the number and types of violation notices we assessed from our sample of 134 notices of violation, for fiscal years 2008 through 2013, for each state and EPA-managed program in our review.

---

[84] According to EPA officials, the information that ultimately gets reported on the 7520-4 form is based on a quarterly calculation of how long the well has been out of compliance; however, according to the officials, EPA only requires state and EPA-managed programs to submit 7520-4 forms to EPA semiannually.

[85] EPA, Office of Drinking Water, *FY 1987 Reporting Requirements—Underground Injection Control Program Guidance (UICP) Guidance #53* (Washington, D.C.: December 1986).

[86] For the purposes of this report, we refer to all written notifications to operators that they are in violation of state or EPA requirements as notices of violation. During our review of seven states, we noted that state or EPA-managed programs may initiate a single enforcement case against an operator for multiple violations, so a single notice of violation may cover more than one violation.

Table 8. Selected State and EPA-Managed Programs' Significant Violation Enforcement Actions from Fiscal Year 2008 through Fiscal Year 2013 and 7520-4 Reporting

State	Operator notification date[a]	First formal action date	Number of days from notification until formal action	Reported on 7520-4	Resolution Date
KY	12/11/2012	6/19/2014	555	Yes	9/8/2014
KY	9/11/2012	10/10/2013	394	Yes	Unresolved
KY	9/27/2012	10/22/2013	390	Yes	Unresolved
KY	9/16/2009	9/8/2010	357	Yes	Unresolved
KY	2/22/2012	11/27/2012	279	Yes	3/26/2014
KY	12/1/2012	8/28/2013	270	Yes	11/4/2013
KY	8/27/2013	3/7/2014	192	Yes	8/26/2014
OH	6/22/2004	3/16/2009	1,728	No	12/21/2009
OH	6/11/2008	12/12/2011	1,309	No	Unresolved
PA	8/13/2010	6/6/2011[b]	297	No	8/27/2012
PA	2/17/2009	7/1/2009	134	No	8/18/2009
PA	8/19/2011	12/22/2011	125	No	4/19/2012
TX	1/24/2007	2/9/2011	1,477	No	12/29/2014
TX	7/28/2008	10/19/2011	1,178	No	11/18/2013
TX	10/5/2009	7/12/2012	1,011	No	8/31/2013
TX	6/6/2008	10/8/2010	854	No	4/30/2014
TX	11/7/08	1/14/2011	798	No	Unresolved
TX	8/3/2006	5/12/2008	648	No	5/25/2012
TX	11/3/2010	4/5/2012	519	No	11/27/2012
TX	11/23/2010	4/20/2012	514	No	2/28/2014
TX	2/13/2008	5/28/2009	470	No	3/31/2012
TX	3/23/2011	4/30/2012	404	No	5/31/2013
TX	12/10/2009	12/15/2010	370	No	4/30/2012
TX	12/10/2009	12/15/2010	370	No	4/30/2012
TX	10/7/2010	9/14/2011	342	No	Unresolved

TX	4/10/2007	12/18/2007	252	No	8/31/2012
TX	9/14/2009	5/21/2010	249	No	5/9/2011
TX	2/11/2009	8/31/2009	201	No	10/31/2014
TX	9/28/2007	2/8/2008	133	No	12/29/2014
KY	12/11/2012	6/19/2014	555	Yes	9/8/2014

Legend: EPA = Environmental Protection Agency; KY = Kentucky, OH = Ohio, PA = Pennsylvania, TX = Texas. Source: GAO analysis of EPA and selected state data. | GAO-16-281

Notes: We chose not to include California in our detailed analysis of inspection and enforcement information from the states because EPA determined in July 2014 that California's program was not in compliance with state and EPA requirements.

The results of our review of inspection and enforcement information reflect the seven remaining state and EPA-managed programs we reviewed. Kentucky and Pennsylvania are EPA-managed programs.

[a]The operator notification date is the first documented date that the operator was notified of the violation.

[b]In reviewing a draft of this report, EPA stated that this date did not account for the earlier conclusion of a criminal case. The file we reviewed did not include documentation of an earlier criminal case, however.

To establish which of those 93 violations should have been reported on the 7520-4 form, we used EPA's 1987 Strategy and 1986 Program Reporting guidance, which call for state and EPA-managed programs to report information on significant violations that were not resolved within 90 days of discovery and also did not have a formal enforcement action taken against the well operator.

To determine the 90-day allowable time frame, we calculated the number of days between the date the operator was notified of the violation and the date a formal enforcement action was taken,[87] and found that 29 significant violations had gone longer than 90 days without formal enforcement action and should have been reported on the 7520-4 form (see *Table 8*). We then compared the results of our calculation to 7520-4 forms we obtained from EPA for fiscal years 2008 through 2013, and found that 7 of the 29 were reported by the respective program.

[87] EPA's 1987 *Strategy* calls for state or EPA-managed programs to take action within 90 days of identifying the significant violation. GAO calculated this time frame using the number of days from the time the operator was notified of the significant violation to when the state or EPA-managed program took formal enforcement action, as defined by the *Strategy*. Using this calculation provides the most conservative estimate of the time that a program took to determine if the violation should have been reported to EPA on the state's appropriate 7520 form.

Appendix VI: Information on Sample Enforcement Cases from Selected State and EPA-Managed Programs

Table 9. Summary of Enforcement Case Information in Selected State and EPA-Managed Programs We Reviewed

State	Violation type[a]	Significant violation	Operator notification date[b]	Date formal action started	Formal action(s) taken[c]	Date violation resolved	Monetary penalty
CO	Failure to plug	No	2/19/2010	8/2/2010	Admin	Unresolved[d]	No
CO	Mechanical integrity	No	6/17/2009	N/A	None	4/2/2010	No
CO	Mechanical integrity	No	8/30/2010	N/A	None	4/26/2011	No
CO	Mechanical integrity	No	7/7/2011	N/A	None	7/16/2013	No
CO	Reporting	No	3/18/2009	N/A	None	8/15/2013	No
CO	Injection over pressure; reporting	No	10/23/2012	2/11/2013	Consent	11/28/2012	Yes
CO	Mechanical integrity	No	4/17/2008	N/A	None	1/20/2010	No
CO	Mechanical integrity	No	1/31/2012	N/A	None	6/12/2013	No
CO	Reporting	No	3/18/2009	N/A	None	8/15/2013	No
CO	Mechanical integrity	No	6/4/2008	N/A	None	7/10/2009	No
CO	Mechanical integrity	No	4/17/2008	N/A	None	3/18/2009	No
CO	Mechanical integrity	No	8/30/2010	N/A	None	4/20/2011	No
CO	Mechanical integrity	No	11/20/2013	12/16/2013	Consent	1/27/2014	Yes
CO	Unauthorized injection	Yes	4/22/2013	7/8/2013	Consent	4/22/2013	Yes
KY	Unauthorized injection	Yes	9/11/2012	3/5/2014	Admin, Consent, Criminal	Unresolved	Yes
KY	Unauthorized injection	Yes	9/27/2012	10/22/2013	Consent	Unresolved	Yes

State	Violation type[a]	Significant violation	Operator notification date[b]	Date formal action started	Formal action(s) taken[c]	Date violation resolved	Monetary penalty
KY	Unauthorized injection	Yes	9/16/2009	9/8/2010	Consent	Unresolved	Yes
KY	Reporting	No	4/26/2010	1/19/2011	Consent	1/3/2011	Yes
KY	Unauthorized injection	Yes	3/25/2014	4/17/2014	Admin, Consent	1/7/2014	No
KY	Unauthorized injection	Yes	5/21/2009	5/21/2009	Admin, Consent, Criminal	2/18/2013	Yes
KY	Unauthorized injection; mechanical integrity; failure to plug	Yes	2/22/2012	11/27/2012	Consent	3/26/2014	Yes
KY	Pollution	Yes	12/11/2012	6/19/2014	Consent	9/8/2014	No
KY	Mechanical integrity	Yes	2/29/2008	5/22/2008	Admin, Consent	9/9/2009	Yes
KY	Mechanical integrity	Yes	2/29/2008	5/22/2008	Admin, Consent	9/9/2009	Yes
KY	Mechanical integrity	Yes	2/29/2008	5/22/2008	Admin, Consent	9/9/2009	Yes
KY	Mechanical integrity	Yes	2/29/2008	5/22/2008	Admin, Consent	9/9/2009	Yes
KY	Mechanical integrity	Yes	2/29/2008	5/22/2008	Admin, Consent	9/9/2009	Yes
KY	Mechanical integrity	Yes	2/29/2008	5/22/2008	Admin, Consent	9/9/2009	Yes
KY	Mechanical integrity	Yes	2/29/2008	5/22/2008	Admin, Consent	9/9/2009	Yes
KY	Mechanical integrity	Yes	2/29/2008	5/22/2008	Admin, Consent	9/9/2009	Yes
KY	Mechanical integrity	Yes	6/13/2008	6/13/2008	Consent	9/9/2009	No
KY	Unauthorized injection	Yes	6/22/2009	6/22/2009	Admin	9/14/2010	No
KY	Unauthorized injection	Yes	4/6/2009	4/6/2009	Admin, Consent	4/6/2010	Yes
KY	Unauthorized injection; mechanical integrity	Yes	8/27/2013	3/7/2014	Admin	8/26/2014	No
KY	Unauthorized injection	Yes	8/8/2013	8/8/2013	Admin, Consent, Civil, Criminal	2/24/2014	Yes
KY	Unauthorized injection	Yes	8/8/2013	8/8/2013	Admin, Civil, Criminal	1/14/2014	No

Table 9. (Continued).

State	Violation type[a]	Significant violation	Operator notification date[b]	Date formal action started	Formal action(s) taken[c]	Date violation resolved	Monetary penalty
KY	Unauthorized injection	Yes	12/18/2012	12/18/2012	Admin, Consent	5/16/2013	Yes
KY	Unauthorized injection	Yes	8/17/2011	8/17/2011	Admin, Consent	12/22/2011	Yes
KY	Unauthorized injection	Yes	11/10/2009	11/10/2009	Admin	2/16/2010	No
KY	Unauthorized injection	Yes	2/25/2011	11/15/2011	Consent	3/16/2011	Yes
KY	Unauthorized injection	Yes	8/17/2010	8/17/2010	Admin, Criminal	8/30/2010	No
KY	Unauthorized injection	Yes	6/2/2009	6/2/2009	Admin	6/9/2009	Yes
KY	Unauthorized injection	Yes	6/7/2010	6/7/2010	Admin	6/12/2010	No
KY	Injection over pressure	Yes	12/1/2012	8/28/2013	Consent	11/4/2013	Yes
ND	Mechanical integrity	No	Missing data	N/A	None	11/5/2009	No
ND	Mechanical integrity	No	12/21/2008	N/A	None	4/23/2009	No
ND	Mechanical integrity; reporting	No	4/1/2008	N/A	None	7/16/2008	No
ND	Mechanical integrity	No	11/19/2010	N/A	None	3/13/2012	No
ND	Mechanical integrity	No	3/10/2008	N/A	None	9/17/2008	No
ND	Mechanical integrity	No	6/2/2009	N/A	None	11/16/2009	No
ND	Mechanical integrity	No	3/25/2010	N/A	None	11/9/2010	No
ND	Mechanical integrity	No	11/17/2008	N/A	None	7/20/2009	No
ND	Mechanical integrity	No	9/17/2008	N/A	None	6/15/2009	No
ND	Injection over pressure	No	7/9/2009	N/A	None	07/27/2009	No
ND	Mechanical integrity	No	8/23/2011	N/A	None	10/2/2012	No
ND	Mechanical integrity	No	7/22/2011	N/A	None	11/16/2012	No
ND	Mechanical integrity	No	9/17/2008	N/A	None	2/11/2010	No
ND	Mechanical integrity	No	10/7/2013	N/A	None	8/3/2015	No

State	Violation type[a]	Significant violation	Operator notification date[b]	Date formal action started	Formal action(s) taken[c]	Date violation resolved	Monetary penalty
ND	Mechanical integrity; failure to plug	No	5/22/2013	N/A	None	Unresolved	No
ND	Mechanical integrity; failure to plug	No	10/10/2012	N/A	None	9/10/2015	No
ND	Mechanical integrity; failure to plug	No	9/6/2012	N/A	None	12/31/2014	No
ND	Mechanical integrity; failure to plug	No	9/30/2011	N/A	None	9/2/2014	No
ND	Unauthorized injection; mechanical integrity	No	2/3/2012	11/16/2012	Criminal	2/16/2012	Yes
ND	Mechanical integrity	No	7/30/2010	N/A	None	7/15/2015	No
OH	Failure to plug	Yes	6/11/2008	12/12/2011	Admin, Civil, Criminal	Unresolved	In Process
OH[d]	Failure to plug	Yes	6/22/2004	3/16/2009	Admin	12/21/2009	No
OH[d]	Mechanical integrity	Yes	3/16/2009	3/16/2009	Admin	6/25/2010	No
OH	Pollution	No	5/12/2011	6/21/2011	Admin	11/28/2012	Yes
OH	Pollution	No	5/24/2011	6/8/2011	Admin	2/16/2012	No
OH	Pollution	No	8/8/2011	8/9/2011	Admin	8/16/2011	No
OH[e]	Pollution	No	2/6/2013	2/6/2013	Admin, Civil, Criminal	2/6/2013	No
OH	Mechanical integrity	Yes	6/18/2008	6/19/2008	Admin	9/8/2009	No
OH	Mechanical integrity; pollution	Yes	5/4/2012	5/10/2012	Admin	9/4/2012	No
OH	Mechanical integrity	Yes	3/19/2009	3/19/2009	Admin	7/16/2009	No
OH	Mechanical integrity	Yes	12/4/2008	12/9/2008	Admin	3/5/2009	No
OH	Mechanical integrity	Yes	11/17/2009	11/19/2009	Admin	1/19/2010	No
OH	Mechanical integrity	Yes	12/29/2008	12/29/2008	Admin	2/20/2009	No

Table 9. (Continued).

State	Violation type[a]	Significant violation	Operator notification date[b]	Date formal action started	Formal action(s) taken[c]	Date violation resolved	Monetary penalty
OH	Mechanical integrity	Yes	12/29/2008	12/29/2008	Admin	2/17/2009	No
OH	Mechanical integrity	Yes	2/2/2010	2/5/2010	Admin	3/23/2010	No
OH	Mechanical integrity	Yes	4/4/2011	4/4/2011	Admin	5/23/2011	No
OH	Mechanical integrity; injection above pressure	Yes	3/25/2009	3/26/2009	Admin	4/20/2009	No
OH	Mechanical integrity	Yes	11/17/2011	11/21/2011	Admin	12/12/2011	No
OH	Mechanical integrity	Yes	10/11/2012	10/22/2012	Admin	10/26/2012	No
OH	Mechanical integrity	Yes	11/1/2010	11/1/2010	Admin	11/16/2010	No
OH	Mechanical integrity	Yes	4/1/2008	4/1/2008	Admin	4/8/2008	No
OH	Mechanical integrity; Pollution	Yes	10/12/2007	10/15/2007	Admin	10/17/2007	No
OK	Unauthorized injection; Mechanical integrity; Reporting	Yes	10/10/2013	10/10/2013[f]	Pending[g]	Unresolved	Pending
OK	Unauthorized injection; mechanical integrity; reporting	Yes	3/16/2012	3/16/2012	Admin	9/21/15	No
OK	Unauthorized injection; mechanical integrity; pollution; reporting	Yes	11/19/2012	11/19/2012	Admin	11/4/2014	Yes
OK	Unauthorized injection; mechanical integrity; reporting	Yes	6/25/2012	6/25/2012	Admin	4/23/2013	No

State	Violation type[a]	Significant violation	Operator notification date[b]	Date formal action started	Formal action(s) taken[c]	Date violation resolved	Monetary penalty
OK	Unauthorized injection; mechanical integrity; reporting	Yes	12/30/2013	12/30/2013	Admin	10/7/2014	Yes
OK	Unauthorized injection; mechanical integrity; reporting	Yes	6/17/2013	6/17/2013	Admin	3/20/2014	Yes
OK	Unauthorized injection; reporting	Yes	1/6/2014	1/6/2014	Admin	10/2/2014	No
OK	Unauthorized injection; mechanical integrity; reporting	Yes	7/12/2013	7/12/2013	Admin	4/1/2014	Yes
OK	Mechanical integrity	Yes	6/8/2012	6/8/2012	Admin	1/15/2013	Yes
OK	Unauthorized injection; mechanical integrity; reporting	Yes	6/7/2012	6/7/2012	Admin	1/10/2013	Yes
OK	Mechanical integrity; violation of formal order; reporting	Yes	5/9/2011	5/9/2011	Admin	11/15/2011	Yes
OK	Mechanical integrity	No	11/14/2013	11/14/2013	Admin	10/1/2015	Yes
OK	Mechanical integrity	No	11/14/2013	11/14/2013	Admin	7/14/2014	No
OK	Mechanical integrity	Yes	4/26/2013	4/26/2013	Admin	10/22/2013	Yes
OK	Unauthorized injection; mechanical integrity; reporting	Yes	11/5/2013	11/5/2013	Admin	3/18/2014	Yes
OK	Unauthorized injection; mechanical integrity; reporting	Yes	6/17/2013	6/17/2013	Admin	10/10/2013	Yes

Table 9. (Continued).

State	Violation type[a]	Significant violation	Operator notification date[b]	Date formal action started	Formal action(s) taken[c]	Date violation resolved	Monetary penalty
OK	Unauthorized injection; mechanical integrity; reporting	Yes	4/30/2013	4/30/2013	Admin	8/22/2013	Yes
OK	Unauthorized injection; mechanical integrity; reporting	Yes	3/13/2012	3/13/2012	Admin	6/28/2012	No
OK	Unauthorized injection; mechanical integrity; reporting	Yes	5/9/2011	5/9/2011	Admin	8/18/2011	Yes
OK	Unauthorized injection; mechanical integrity; reporting	Yes	6/23/2010	6/23/2010	Admin	9/30/2010	Yes
OK	Unauthorized injection; mechanical integrity; injection over pressure; reporting	Yes	3/8/2012	3/8/2012	Admin	5/31/2012	Yes
OK	Unauthorized injection; mechanical integrity; reporting	Yes	3/31/2011	3/31/2011	Admin	5/10/2011	Yes
PA	Mechanical integrity	No	5/21/2008	5/25/2010	Admin, Consent	6/17/2014	No
PA	Unauthorized injection	Yes	5/12/2011	5/12/2011	Admin	4/1/2014	No
PA	Unauthorized injection	Yes	8/13/2010	6/6/2011[h]	Admin, Consent, Criminal	8/27/2012	Yes
PA	Mechanical integrity	Yes	8/19/2011	12/22/2011	Admin, Consent	4/19/2012	Yes

State	Violation type[a]	Significant violation	Operator notification date[b]	Date formal action started	Formal action(s) taken[c]	Date violation resolved	Monetary penalty
PA	Unauthorized injection	Yes	2/17/2009	7/1/2009	Admin, Consent	8/18/2009	Yes
PA	Unauthorized injection	Yes	9/24/2010	9/24/2010	Admin, Consent	9/27/2010	Yes
TX	Mechanical integrity; failure to plug	Yes	11/7/2008	1/14/2011	Admin, Civil	Unresolved	Yes
TX	Mechanical integrity; failure to plug	Yes	10/7/2010	9/14/2011	Admin, Civil	Unresolved	Yes
TX	Unauthorized injection; failure to plug	Yes	2/11/2009	8/31/2009	Admin, Civil	10/31/2014	Yes
TX	Mechanical integrity; pollution	Yes	6/8/2009	8/18/2009	Admin, Civil	8/25/2011	Yes
TX	Mechanical integrity; failure to plug	Yes	1/24/2007	2/9/2011	Admin, Civil	12/29/2014	Yes
TX	Mechanical integrity; failure to plug	Yes	9/28/2007	2/8/2008	Admin, Civil	12/29/2014	Yes
TX	Mechanical integrity; failure to plug	Yes	6/6/2008	10/8/2010	Admin, Civil	4/30/2014	Yes
TX	Failure to plug	Yes	8/3/2006	5/12/2008	Admin, Civil	5/25/2012	Yes
TX	Mechanical integrity; failure to plug	Yes	4/10/2007	12/18/2007	Admin, Consent, Civil	8/31/2012	Yes
TX	Mechanical integrity	Yes	7/28/2008	10/19/2011	Admin, Civil	11/18/2013	Yes
TX	Mechanical integrity; failure to plug	Yes	7/15/2009	10/8/2009	Admin, Civil	1/15/2014	Yes
TX	Mechanical integrity; failure to plug;	Yes	2/13/2008	5/28/2009	Admin, Civil	3/31/2012	Yes
TX	Mechanical integrity; failure to plug	Yes	10/5/2009	7/12/2012	Admin, Civil	8/31/2013	Yes

Table 9. (Continued).

State	Violation type[a]	Significant violation	Operator notification date[b]	Date formal action started	Formal action(s) taken[c]	Date violation resolved	Monetary penalty
TX	Mechanical integrity; failure to plug	Yes	11/23/2010	4/20/2012	Admin, Civil	2/28/2014	Yes
TX	Failure to plug	Yes	12/10/2009	12/15/2010	Admin, Civil	4/30/2012	Yes
TX	Mechanical integrity; failure to plug	Yes	12/10/2009	12/15/2010	Admin, Civil	4/30/2012	Yes
TX	Mechanical integrity; failure to plug	Yes	3/23/2011	4/30/2012	Admin, Civil	5/31/2013	Yes
TX	Mechanical integrity; unauthorized injection	Yes	11/3/2010	4/5/2012	Admin, Consent	11/27/2012	Yes
TX	Mechanical integrity	Yes	9/14/2009	5/21/2010	Admin, Consent, Civil	5/9/2011	Yes
TX	Mechanical integrity; failure to plug	Yes	10/9/2012	10/9/2012	Admin, Consent	12/31/2013	Yes

Legend: EPA = Environmental Protection Agency; CO = Colorado; KY = Kentucky; ND = North Dakota; OH = Ohio; OK = Oklahoma; PA = Pennsylvania; TX = Texas; N/A = not applicable, Admin = administrative order; Consent = consent agreement; Civil = civil judicial action; Criminal = criminal judicial action.

Source: GAO analysis of EPA and selected state data. | GAO-16-281

Notes: We chose not to include California in our detailed analysis of inspection and enforcement information from the states because EPA determined in July 2014 that California's program was not in compliance with state and EPA requirements. The results of our review of inspection and enforcement information reflect the seven remaining state and EPA-managed programs we reviewed. Kentucky and Pennsylvania are EPA-managed programs.

[a]This table includes only the most serious violations from each enforcement case.

[b]The operator notification date is the first documented date that the operator was notified of the violation.

[c]Formal actions taken can include administrative orders—both unilateral (admin) and negotiated (consent); civil judicial referrals (civil); and criminal judicial referrals (criminal). The ordering of the actions listed does not represent the order in which those actions took place in the enforcement process.

[d]According to Colorado officials, the well in violation is scheduled to be plugged by the state in 2016.

[e]Ohio's program requested EPA's assistance in this enforcement case. According to Ohio officials, with EPA's assistance, the violation was resolved.

[f]According to Ohio program officials, EPA pursued criminal judicial action, and Ohio's Environmental Protection Agency pursued civil judicial action, against the operator.

[g]According to Oklahoma program officials, after discovering a violation, the program will issue a citation ordering the well operator to appear before an administrative law judge. The citation does not conform to EPA's definition of a formal enforcement action. However, because in many cases Oklahoma did not issue a formal order for 1 to 2 more years following the citation, and because those orders tended to be final orders memorializing the fact that the violation had already been resolved, we used the citation date as the date that formal enforcement action began.

[h]The well operator went out of business prior to Oklahoma issuing an administrative order or consent order.

[i]In reviewing a draft of this report, EPA stated that this date did not account for the earlier conclusion of a criminal case. The file we reviewed did not include documentation of an earlier criminal case, however.

Appendix VII: Comments from the Environmental Protection Agency

UNITED STATES ENVIRONMENTAL PROTECTION AGENCY
WASHINGTON, D.C. 20460

Mr. J. Alfredo Gomez
Director
Natural Resources and Environment
U.S. Government Accountability Office
Washington, DC 20548

FEB 10 2016

Dear Mr. Gomez:

Thank you for the opportunity to review and comment on the Government Accountability Office's Draft Report GA0-16-281, "DRINKING WATER: EPA Needs to Collect Information and Consistently Conduct Activities to Protect Underground Sources of Drinking Water." In this draft report, GAO reviewed the U.S. Environmental Protection Agency's oversight of state and EPA Class II Underground Injection Control programs (i.e., pertaining to oil and gas-related injection fluids). GAO evaluated the EPA's collection of inspection and enforcement records to determine whether the agency is conducting the oversight activities needed to assess the Class II UIC program's protection of underground sources of drinking water. The purpose of this letter is to provide the EPA's response to your findings, conclusions and recommendations. We also provide technical comments in a separate document.

The agency has an important role in oversight of the UIC programs administered by the states and directly implemented by EPA Regions, and is continually working to improve its oversight to ensure protection of underground sources of drinking water. Oversight of state UIC programs is the joint responsibility of several EPA offices, including the EPA Headquarters' Office of Water and Office of Enforcement and Compliance Assurance as well as the EPA Regional water program and enforcement offices. In addition to their state program oversight responsibilities, the EPA Regional offices are also responsible for direct implementation of the UIC program in states that do not have primary enforcement responsibility. Additionally, although all state UIC programs have the goal of effectively protecting USDWs, Class II injection regulations differ by state according to their geology, hydrology and injection needs. Therefore, each state program may have requirements and solutions tailored to its individual circumstances. While GAO's investigation focused on UIC activities associated with oil and gas, the agency's efforts encompass all six well classes, including more than 650,000 injection wells, which are overseen by 59 state agencies and the 10 EPA Regional offices. Any discussion of oversight responsibilities and resource needs will be relevant to the entire UIC program.

The agency acknowledges that oversight and data management are two long-term challenges, especially in light of significant variations in state Class II programs, as allowed by the Safe Drinking Water Act. The agency is in the process of modernizing EPA's well inventory and compliance data collection and focusing its oversight resources on priority risks to public health. The agency's oversight helps ensure that the states, tribes and territories, who have been granted primacy under the SDWA for the UIC program, are implementing their programs consistent with the statute and relevant state and federal UIC regulations. The statute and regulations require that Class II UIC programs permit underground injection wells in a manner that is protective of USDWs. As noted above, the EPA Regional water and enforcement programs conduct the primary oversight of state U IC programs. In addition to a significant level of informal coordination, the EPA's state UIC program oversight consists of the following efforts, conducted annually:

- Setting state/regional targets, tracking progress and reporting for EPA's National Water Program Activity Measures;

- Collecting, reviewing and confirming state UIC Annual Program Reports and well inventories;
- Collecting, reviewing and compiling compliance and enforcement reports (EPA 7520 forms);
- Developing a grant workplan with each state to target UIC program needs with funds and activities;
- Collecting and reviewing U IC program grant expenditure reporting; and
- Conducting an informal evaluation of state/tribal UIC program performance against workplan objectives and targets during the grant year to identify and resolve any performance issues.

EPA Regional UIC programs work with states to improve their U IC programs with the information collected annually through the above activities, as well as with the knowledge developed through informal exchanges, such as compliance assistance and program and policy updates. The EPA Regions also provide training, technical support and guidance to states on issues of national significance in coordination with EPA Headquarters. States develop and prioritize UIC annual activities with agency input to optimize UIC program implementation. This approach allows for frequent interaction between states and the EPA as well as the flexibility for the agency to target oversight efforts in quick response to emerging UIC issues.

As needed, Regional EPA UIC programs conduct more extensive, formal evaluations of state UIC programs. Some evaluative efforts are broad in scope, covering the entire program. Other evaluations are more targeted toward specific concerns such as injection induced seismicity. EPA Headquarters may participate actively in either broad or targeted evaluations. Both types of evaluations require extensive planning and coordination between the states and the agency. Some Regions incorporate compliance and enforcement into a broader program review, while others evaluate them separately, depending on how the regional water and enforcement offices are organized. Once the agency completes the review and analysis of information in concert with the state agency, the EPA prepares a document that includes recommendations, as appropriate, to help the state more effectively implement its UIC program. All of these efforts are designed to ensure regulatory safeguards are in place, and to improve implementation and understanding of state and EPA UIC programs across the nation.

EPA agrees with GAO that having high quality data is an important part of these oversight activities. Collection of well-specific data continues to be a key goal for the EPA. Al though we do not think that it makes sense now to require the states to submit well-specific data, we will continue to work with our state partners to improve both the collection and the quality of the data currently required and to expand EPA's access to additional state data. The agency also agrees that continued improvement in collection and consistency of data via the 7520-4 form would be valuable to more effective oversight. The agency created a standard operating procedure for 7520 forms in response to GAO's 2014[88] report to improve consistency in reporting. The agency will provide additional materials to states and Regions to clarify how they should complete the 7520-4 form to further ensure the information is consistent and complete.

The agency disagrees with GAO's assessment that the EPA is deficient in its duties to maintain aquifer exemption records. The draft report presents incomplete infom1ation as to which materials are held at the EPA Headquarters and Regional levels and the roles and objectives that EPA Headquarters and the Regions play regarding aquifer exemptions and the use of data. The GAO asserts that, "EPA does not have sufficient information on aquifer exemptions to oversee state and EPA-managed programs and assess whet her programs are protecting underground sources of drinking water." This statement is incorrect. The EPA Regions hold the most current and comprehensive data on aquifer exemption requests and approvals for their states to support their review and approval. EPA Headquarters is collecting selected data from the Regions on

[88] *Drinking Water: EPA Program to Pr0lect Underground Sources from Injection of Fluids Associated with Oil and Gas Production Needs Improvement,* GA0-14-555 (Washington, DC: June 27, 2014).

approved aquifer exemptions throughout the nation to develop a national da ta set that it plans to release publicly in 2016.

While the EPA agrees that oversight is an important aspect of ensuring an effective UIC program, a workforce analysis is not necessary to better assess the resources needed to oversee the implementation of the UIC Class II program. Rather, we will assess and, if necessary, expand our current evaluation of UIC program oversight to include elements of inspection and enforcement. Once the evaluation is complete, we would look to improve the effectiveness of state and EPA UIC oversight if needed.

GAO Recommendations and EPA Response

To help ensure protection of underground sources of drinking water from the injection of fluids associated with domestic oil and gas production, GAO recommends that EPA take four actions.

Recommendation 1

Require and collect well-specific data 011 inspections from state and EPA- managed programs, including when the wells were inspected, the types of inspections conducted, and the results of the inspections in order to track progress toward state and EPA-managed annual inspection goals.

EPA Response

The agency's goal is to obtain high quality data to understand program activities at the well-specific level, and we are working toward establishing a complete, regularly updated data set. While we agree that this infom1ation is important, we are mindful of the need to think carefully about instituting new reporting requirements. Such requirements could necessitate state UIC programs to direct limited resources to develop new reporting structures and additional data input, potentially distracting from the inspection work in the states. The existing approach was developed with the input from state regulators who are closest to the ground, and any change in approach should continue to take advantage of state knowledge and expertise. Because we agree that this type of information is valuable, we will continue to explore options to improve collection of data and use of existing tools in accordance with reporting requirements and work to expand the agency's access to well-specific inspection and enforcement data. If the agency were to consider further changes to the 7520 reporting forms in the future, we would continue to work with state co-regulators to make sure that reporting is as efficient and useful as possible.

The current mechanism for required reporting from the states is the set of 7520 forms for summary information. In its previous report, GAO identified room for improvement in this collection activity and the agency has taken steps to address these gaps, including developing standard operating procedures for submission and review of the data forms and revising instructions to increase consistency in reporting the data to the EPA. As a result of these efforts, we expect improved information from future submissions to allow for better assessment of program activities.

We also continue to increase the inventory of well-specific data in our National UIC Database. We have made data collection from EPA-managed programs a priority and currently have well-specific inspection data for EPA-managed programs in all EPA Regions. Additionally, we continue to work with state programs to encourage them to share well-specific inspection data. While GAO correctly notes that the only non-EPA Class II programs currently using the database as their official database of record are the State of Montana and the Navajo Nation, there are other states working towards e-reporting status and we welcome and encourage further participation. Finally, we continue to work with the Ground Water Protection Council and the Department of Energy as they develop a National Oil and Gas Gateway for well-specific state data. Although the Gateway does not include inspection data, its well-specific inventory data is valuable to the agency's ability to understand and help improve states' Class II primacy programs.

Recommendation 2

Complete the aquifer exemption database and a way to update it to provide EPA headquarters and Regions with sufficient information of aquifer exemptions to oversee state and EPA-managed programs.

EPA Response

EPA Headquarters is engaged in an effort to collect, standardize and centralize aquifer exemption boundaries and other key data from the EPA Regional offices. The agency initiated this effort to better understand the number, the locations and the nature and quality of aquifers exempted by the UIC program. Our intent is also to share key information with the public to improve transparency. The EPA Regional offices hold the most comprehensive and current data on aquifer exemptions as they conduct the final review of exemption requests and must approve all exemptions.

We expect to release a public data set by the end of 2016, which will include data current through 2015, with the exception of Region 9's data for the State of California. The public data set will show the approved aquifer exemption boundaries in two dimensions as well as information such as depth of injection, surrounding geology and injectate characteristics. We anticipate adding Region 9's aquifer exemption data for California as the Region works with the state to clarify the boundaries of the agency's historic approvals and takes action on the state's requests for new exemptions. EPA plans to update the central public data set annually, and the Regions will continue to hold the most current data. Public notification in the local area, including the exemption request documentation and the opportunity for public comment, accompanies all new requests for aquifer exemptions. The public will retain the ability to obtain information about aquifer exemptions from the EPA Regions between annual updates of the national data set.

Recommendation 3

Clarify guidance on what data should be reported on the 7520-4 form to help ensure that the data collected are complete and consistent across state and EPA-managed programs and to provide the information EPA needs to assess whether to take enforcement action.

EPA Response

The 7520-4 form is used to track specific wells in significant non-compliance, which a primacy program has not addressed with an enforcement action or that have not returned to compliance for two or more quarters. The agency agrees that the form is a tool for obtaining important information used in assessing enforcement activities and that providing guidance on the 7520-4 could be valuable to improve the quality of information the agency receives. The current 7520 standard operating procedures that the EPA created in response to GAO's previous report reminds reviewers that wells in significant non-compliance for two or more quarters should remain listed on the 7520-4 until the issue is resolved. This procedure was established to address an area where reporting has been inconsistent. The agency will provide further materials to U IC data submitters, to improve completeness and consistency of the data a U!C program reports on form 7520-4, within six months of GAO's final report.

Recommendation 4

EPA should conduct a workforce analysis to identify the resources EPA needs to conduct program oversight.

EPA Response

The EPA agrees that oversight is an important aspect of ensuring an effective UIC program; however, a workforce analysis is not necessary to better assess the resources needed to oversee the implementation of the UIC Class II program. In response to GAO's previous report, we are working with the UIC program managers to evaluate the effectiveness of our current oversight activities. The EPA believes the best approach is to expand this evaluation to include elements of inspection and enforcement. Once the evaluation is complete, we would look to improve the effectiveness of state and EPA UIC oversight if needed. For example, piloting a project that would explore the potential to assure program implementation by use of remote approaches, such as data collection, data analysis, targeting and priority ranking, and public transparency, may provide a viable option for increased oversight.

The agency appreciates the efforts that GAO expended in conducting this review. The agency generally agrees with GAO's analysis and findings on the UIC Class II program as reflected in our responses. Thank you for the opportunity to provide comments on the draft report. The EPA looks forward to working collaboratively with GAO to continue to protect drinking water sources and public health. If you have any questions, please contact Holly Green at (202) 566-0651 or Gwendolyn Spriggs at (202) 564-2439.

Sincerely,

Cynthia Giles
Assistant Administrator
Office of Enforcement and Compliance Assurance

Joel Beauvais
Deputy Assistant Administrator
Office of Water

Enclosure

cc: EPA GAO Liaison Team
Peter Grevatt (OGWDW)
Betsy Smidinger (OC)
Susan Shinkman (OCE)
Anita Thompkins (OGWDW)
Ronald Bergman (OGWDW)
Holly G reen (OGWDW)
Carrie Wehling (OGC)

Appendix VIII: GAO Contact and Staff Acknowledgments

GAO Contact

J. Alfredo Gómez, (202) 512-3841 or gomezj@gao.gov

Staff Acknowledgments

In addition to the contact named above, Susan Lott (Assistant Director), Mark Braza, Antoinette Capaccio, John Delicath, John Hocker, Rich Johnson, Micah McMillan, Maria Stattel, Kiki Theodoropoulos and Breanna Trexler made key contributions to this report.

Appendix IX: Accessible Data

Agency Comment Letter

Text of Appendix VII: Comments from the Environmental Protection Agency

UNITED STATES ENVIRONMENTAL PROTECTION AGENCY
ASHINGTON, D.C. 20460

Mr. J. Alfredo Gomez

Director

Natural Resources and Environment U.S. Government Accountability Office Washington, DC 20548

FEB 10 2016

Dear Mr. Gomez:

Thank you for the opportunity to review and comment on the Government Accountability Office's Draft Report GA0-16-281, "DRINKING WATER: EPA Needs to Collect Information and Consistently Conduct Activities to Protect Underground Sources of Drinking Water." In this draft report,

GAO reviewed the U.S. Environmental Protection Agency's oversight of state and EPA Class II Underground Injection Control programs (i.e., pertaining to oil and gas-related injection fluids). GAO evaluated the EPA's collection of inspection and enforcement records to determine whether the agency is conducting the oversight activities needed to assess the Class II UIC program's protection of underground sources of drinking water. The purpose of this letter is to provide the EPA's response to your findings, conclusions and recommendations. We also provide technical comments in a separate document.

The agency has an important role in oversight of the UIC programs administered by the states and directly implemented by EPA Regions, and is continually working to improve its oversight to ensure protection of underground sources of drinking water. Oversight of state UIC programs is the joint responsibility of several EPA offices, including the EPA Headquarters ' Office of Water and Office of Enforcement and Compliance Assurance as well as the EPA Regional water program and enforcement offices. In addition to their state program oversight responsibilities, the EPA Regional offices are also responsible for direct implementation of the UIC program in states that do not have primary enforcement responsibility. Additionally, although all state UIC programs have the goal of effectively protecting USDWs, Class II injection regulations differ by state according to their geology, hydrology and injection needs. Therefore, each state program may have requirements and solutions tailored to its individual circumstances. While GAO's investigation focused on UIC activities associated with oil and gas, the agency's efforts encompass all six well classes, including more than 650,000 injection wells, which are overseen by 59 state agencies and the 10 EPA Regional offices. Any discussion of oversight responsibilities and resource needs will be relevant to the entire UIC program.

The agency acknowledges that oversight and data management are two long-tern1challenges, especially in light of significant variations in state Class II programs, as allowed by the Safe Drinking Water Act. The agency is in the process of modernizing EPA's well inventory and compliance data collection and focusing its oversight resources on priority risks to public health. The agency's oversight helps ensure that the states, tribes and territories, who have been granted primacy under the SDWA for the UIC program, are implementing their programs consistent with the statute and relevant state and federal UIC regulations. The statute and regulations require that Class II UIC programs permit underground injection wells in a manner that is protective of USDWs. As noted above, the EPA Regional water and enforcement programs conduct the primary oversight of state UIC programs. In addition to a significant level of informal coordination, the EPA's state UIC program oversight consists of the following efforts, conducted annually:

- Setting state/regional targets, tracking progress and reporting for EPA's National Water Program Activity Measures;
- Collecting, reviewing and confirming state UIC Annual Program Reports and well inventories;
- Collecting, reviewing and compiling compliance and enforcement reports (EPA 7520 forms);
- Developing a grant workplan with each state to target UIC program needs with funds and activities;
- Collecting and reviewing UIC program grant expenditure reporting; and
- Conducting an informal evaluation of state/tribal UIC program performance against workplan objectives and targets during the grant year to identify and resolve any performance issues.

EPA Regional UIC programs work with states to improve their UIC programs with the information collected annually through the above activities, as well as with the knowledge developed through informal exchanges, such as compliance assistance and program and policy updates. The EPA Regions also provide training, technical support and guidance to states on issues of national significance in coordination with EPA Headquarters. States develop and prioritize UIC annual activities with agency input to optimize UJC program implementation. This approach allows for frequent interaction between states and the EPA as well as the flexibility for the agency to target oversight efforts in quick response to emerging UIC issues.

As needed, Regional EPA U IC programs conduct more extensive, formal evaluations of state UIC programs. Some evaluative efforts are broad in scope, covering the entire program. Other evaluations are more targeted toward specific concerns such as injection induced seismicity. EPA Headquarters may participate actively in either broad or targeted evaluations. Both types of evaluations require extensive planning and coordination between the states and the agency. Some Regions incorporate compliance and enforcement into a broader program review, while others evaluate them separately, depending on how the regional water and enforcement offices are organized. Once the agency completes the review and analysis of information in concert with the state agency, the EPA prepares a document that includes recommendations, as appropriate, to help the state more effectively implement its UIC program. All of these efforts are designed to ensure regulatory safeguards are in place, and to improve implementation and understanding of state and EPA UIC programs across the nation.

EPA agrees with GAO that having high quality data is an important part of these oversight activities. Collection of well-specific data continues to be a key goal for the EPA. Although we do not think that it makes sense now to require the states to submit well-specific data, we will continue to work with our state partners to improve both the collection and the quality of the data currently required and to expand EPA 's access to additional state data. The agency also agrees that continued improvement in collection and consistency of data via the 7520-4 form would be valuable to more effective oversight. The agency created a standard operating procedure for 7520 forms in response to GAO's 2014[89] report to improve consistency in reporting. The agency will provide additional materials to states and Regions to clarify how they should complete the 7520-4 form to further ensure the information is consistent and complete.

The agency disagrees with GAO's assessment that the EPA is deficient in its duties to maintain aquifer exemption records. The draft report presents incomplete information as to which materials are held at the EPA Headquarters and Regional levels and the roles and objectives that EPA Headquarters and the Regions play regarding aquifer exemptions and the use of data. The GAO asserts that, "EPA does not have sufficient information on aquifer exemptions to oversee state and EPA-managed programs and assess whether programs are protecting underground sources of drinking water." This statement is incorrect. The EPA Regions hold the most current and comprehensive data on aquifer exemption requests and approvals for their states to support their review and approval. EPA Headquarters is collecting selected data from the Regions on approved aquifer exemptions throughout the nation to develop a national data set that it plans to release publicly in 2016.

While the EPA agrees that oversight is an important aspect of ensuring an effective UIC program, a workforce analysis is not necessary to better assess the resources needed to oversee the implementation of the UIC Class II program. Rather, we will assess and, if necessary, expand our current evaluation of UIC program oversight to include elements of inspection and enforcement. Once the evaluation is complete, we would look to improve the effectiveness of state and EPA UIC oversight if needed.

GAO Recommendations and EPA Response

To help ensure protection of underground sources of drinking water from the injection of fluids associated with domestic oil and gas production, GAO recommends that EPA take four actions.

---

[89] Drinking Water: EPA Program to Protect Underground Sources from Injection of Fluids Associated with Oil and Gas Production Needs Improvement, GA0-14-555 (Washington, DC: June 27, 20 14).

Recommendation 1

Require and collect well-specific data on inspections from state and EPA managed programs, including when the wells were inspected, the types of inspections conducted, and the results of the inspections in order to track progress toward state and EPA-managed annual inspection goals.

EPA Response

The agency's goal is to obtain high quality data to understand program activities at the well-specific level, and we are working toward establishing a complete, regularly updated data set. While we agree that this information is important, we are mindful of the need to think carefully about instituting new reporting requirements. Such requirements could necessitate state UIC programs to direct limited resources to develop new reporting structures and additional data input, potentially distracting from the inspection work in the states. The existing approach was developed with the input from state regulators who are closest to the ground, and any change in approach should continue to take advantage of state knowledge and expertise. Because we agree that this type of information is valuable, we will continue to explore options to improve collection of data and use of existing tools in accordance with reporting requirements and work to expand the agency's access to well- specific inspection and enforcement data. If the agency were to consider further changes to the 7520 reporting forms in the future, we would continue to work with state co-regulators to make sure that reporting is as efficient and useful as possible.

The current mechanism for required reporting from the states is the set of 7520 forms for summary information. In its previous report, GAO identified room for improvement in this collection activity and the agency has taken steps to address these gaps, including developing standard operating procedures for submission and review of the data forms and revising instructions to increase consistency in reporting the data to the EPA. As a result of these efforts, we expect improved information from future submissions to allow for better assessment of program activities.

We also continue to increase the inventory of well-specific data in our National UIC Database. We have made data collection from EPA-managed programs a priority and currently have well-specific inspection data for EPA-managed programs in all EPA Regions. Additionally, we continue to work with state programs to encourage them to share well-specific inspection data. While GAO correctly notes that the only non-EPA Class II programs currently using the database as their official database of record are the State of Montana and the Navajo Nation, there are other states working towards e-reporting status and we welcome and encourage further participation. Finally, we continue to work with the Ground Water Protection Council and the Department of Energy as they develop a National Oil and Gas Gateway for well-specific state data. Although the Gateway does not include inspection data, its well-specific inventory data is valuable to the agency's ability to understand and help improve states' Class II primacy programs.

Recommendation 2

Complete the aquifer exemption database and a way to update it to provide EPA headquarters and Regions with sufficient information of aquifer exemptions to oversee state and EPA-managed programs.

EPA Response

EPA Headquarters is engaged in an effort to collect, standardize and centralize aquifer exemption boundaries and other key data from the EPA Regional offices. The agency initiated this effort to better understand the number, the locations and the nature and quality of aquifers exempted by the UIC program. Our intent is also to share key information with the public to improve transparency. The EPA Regional offices hold the most comprehensive and current data on aquifer exemptions as they conduct the final review of exemption requests and must approve all exemptions.

We expect to release a public data set by the end of 2016, which will include data current through 2015, with the exception of Region 9's data for the State of California. The public data set will show the approved aquifer exemption boundaries in two dimensions as well as information such as depth of injection, surrounding geology and injectate characteristics. We anticipate adding Region 9's aquifer exemption data for

California as the Region works with the state to clarify the boundaries of the agency's historic approvals and takes action on the state's requests for new exemptions. EPA plans to update the central public data set annually, and the Regions will continue to hold the most current data. Public notification in the local area, including the exemption request documentation and the opportunity for public comment, accompanies all new requests for aquifer exemptions. The public will retain the ability to obtain information about aquifer exemptions from the EPA Regions between annual updates of the national data set.

Recommendation 3

Clarify guidance Oil what data should be reported Oil the 7520-4form to help ensure that the data collected are complete and consistent across state and EPA-managed programs and to provide the information EPA needs to assess whether to take enforcement action.

EPA Response

The 7520-4 form is used to track specific wells in significant non-compliance, which a primacy program has not addressed with an enforcement action or that have not returned to compliance for two or more quarters. The agency agrees that the form is a tool for obtaining important information used in assessing enforcement activities and that providing guidance on the 7520-4 could be valuable to improve the quality of information the agency receives. The current 7520 standard operating procedures that the EPA created in response to GAO's previous report reminds reviewers that wells in significant non-compliance for two or more quarters should remain listed on the 7520-4 until the issue is resolved. This procedure was established to address an area where reporting has been inconsistent. The agency will provide further materials to U IC data submitters, to improve completeness and consistency of the data a UIC program reports on form 7520-4, within six months of GAO's final report.

Recommendation 4

EPA should conduct a workforce analysis to identify the resources EPA needs to conduct program oversight.

EPA Response

The EPA agrees that oversight is an important aspect of ensuring an effective UIC program; however, a workforce analysis is not necessary to better assess the resources needed to oversee the implementation of the U IC Class II program. In response to GAO's previous report, we are working with the UIC program managers to evaluate the effectiveness of our current oversight activities. The EPA believes the best approach is to expand this evaluation to include elements of inspection and enforcement. Once the evaluation is complete, we would look to improve the effectiveness of state and EPA U IC oversight if needed. For example, piloting a project that would explore the potential to assure program implementation by use of remote approaches, such as data collection, data analysis, targeting and priority ranking, and public transparency, may provide a viable option for increased oversight.

The agency appreciates the efforts that GAO expended in conducting this review. The agency generally agrees with GAO's analysis and findings on the UIC Class II program as reflected in our responses. Thank you for the opportunity to provide comments on the draft report. The EPA looks forward to working collaboratively with GAO to continue to protect drinking water sources and public health. If you have any questions, please contact Holly Green at (202) 566-0651 or Gwendolyn Spriggs at (202) 564-2439.

Sincerely,

Cynthia Giles

Assistant Administrator

Office Enforcement and Compliance Assurance

Joel Beauvais

Deputy Assistant Administrator

Office of Water

Enclosure

cc:
EPA GAO Liaison Team
Peter Grevatt (OGWDW)

Betsy Smidinger (OC)

Susan Shinkman (OCE)

Anita Thompkins (OGWDW)

Ronald Bergman (OGWDW)

Holly Green (OGWDW)

Carrie Wehling (OGC)

In: Water Quality
Editors: Edward Saunders et al.
ISBN: 978-1-53614-809-1

Chapter 7

LEAD IN FLINT, MICHIGAN'S DRINKING WATER: FEDERAL REGULATORY ROLE*

Mary Tiemann

In April 2014, the city of Flint, Michigan, stopped purchasing treated water from the city of Detroit and began using the Flint River as its water source without providing corrosion control treatment. (A key source of lead in drinking water is the corrosion of plumbing materials in the distribution system and households.) While some water quality changes were quickly apparent, elevated lead concentrations were identified over a longer period through monitoring conducted by the city and others and detections of elevated blood lead levels in children. On October 1, 2015, city officials urged residents to stop drinking the water. On October 16, Flint reconnected to Detroit's water and advised residents not to use unfiltered tap water. The city and the governor of Michigan each declared a state of emergency. President Obama issued an emergency declaration on January 16, 2016.

* This is an edited, reformatted and augmented version of Congressional Research Service, Publication No. IN10446, dated March 2, 2016.

On January 21, 2016, the Environmental Protection Agency (EPA) issued an emergency order directing the city and state to take immediate actions to address concerns over the city's water system. Requirements include, among others, that the city and state re-optimize corrosion control, post online lead monitoring results and weekly reports, and ensure the city's capacity to operate the system in compliance with federal regulations. EPA's current Flint responses include providing technical assistance for water testing and treatment, conducting water monitoring, and identifying lead service line locations.

FEDERAL REGULATORY ROLE

The Safe Drinking Water Act (SDWA) authorizes EPA to regulate contaminants in water supplies. SDWA regulations establish compliance obligations for public water systems (PWSs) and state agencies to which EPA has delegated primary enforcement responsibility for the PWS supervision program. Forty-nine states, including Michigan, have this authority.

States have first-line responsibilities for administering SDWA programs, providing technical assistance to systems, and compelling systems to comply with regulations. Section 1414 provides that, in addition to overseeing states, when EPA finds that a water system is not in compliance, EPA must notify the state and system and provide assistance to both in order to bring the system into compliance. If, after 30 days, a state does not initiate enforcement action, EPA must do so. Under SDWA's emergency powers (Section 1431), EPA "may take such actions as [the Administrator] may deem necessary" to protect human health when a contaminant in a water system "may present an imminent and substantial endangerment to the health of persons" and state and local authorities have not acted. This authority is available even absent a violation. EPA used this authority to issue the emergency order.

LEAD AND COPPER RULE

EPA regulates lead in drinking water through the 1991 Lead and Copper Rule (LCR), which includes use of a treatment technique (corrosion control) to prevent these metals from leaching into water. The LCR includes a lead "action level" of 15 parts per billion. If more than 10% of tap-water samples exceed the action level, a PWS has not violated the rule, but other requirements—corrosion control, public education, additional water monitoring, and lead service line replacement—can be triggered. A system violates the rule if it does not take the triggered actions. If a system plans to change its water source, the state must be notified.

In October 2015, the Michigan Department of Environmental Quality (MDEQ) announced that it had followed the wrong monitoring protocol for Flint. Reportedly, the city conducted water testing incorrectly. In November 2015, EPA clarified corrosion control requirements for large systems that previously purchased treated water.

EPA initiated an extensive review of the LCR in 2004 after increased lead levels were detected in the District of Columbia's water following a water treatment change. EPA promulgated short-term revisions and clarifications in 2007 and continues working on comprehensive revisions. In December 2015, EPA's National Drinking Water Advisory Council presented recommendations for LCR revisions. EPA plans to propose a revised rule in 2017. Topics being evaluated include sampling protocols, lead service line replacement, corrosion control, and risk communication.

CONGRESSIONAL RESPONSES

Events in Flint have raised questions regarding the adequacy of the LCR and scope of EPA enforcement and notification authorities under SDWA.

Further, these events have focused attention on the state of water infrastructure nationwide and the maintenance and repair challenges many communities face. The Drinking Water State Revolving Fund (DWSRF) program provides assistance to systems for projects needed to comply with SDWA regulations and protect public health. The Consolidated Appropriations Act, 2016 (P.L. 114-113) includes $863.2 million for this program. The President has requested $1.02 billion for FY2017.

In addition to attempts to amend S. 2012, legislation has been introduced in response to the Flint crisis. House-passed H.R. 4470 would require systems to notify customers, the state, and EPA of lead action level exceedances and require EPA to notify the public if the system or state has not done so within 24 hours after EPA receives notification. EPA would be required to develop a plan for improving information sharing and public communication. S. 2466 would also authorize EPA to notify the public of action level exceedances. S. 2587 would direct EPA to revise the LCR. H.R. 4438 seeks $1 billion in emergency appropriations for a grant to Michigan to replace water infrastructure. To aid Flint residents, H.R. 4479 would authorize infrastructure grants; increase DWSRF loan forgiveness; authorize assistance under health, education, and other programs; and require establishment of the Center on Excellence on Lead Exposure, an advisory committee, and health registry. S. 2579 would require disclosure of any lead level exceedance within 15 days, provide water infrastructure loan guarantees and DWSRF capitalization grants (and increased loan forgiveness), fund public health programs, and require the establishment of a lead exposure registry and advisory committee. H.R. 4653 would direct EPA to establish a lead service line replacement grant program.

In: Water Quality
Editors: Edward Saunders et al.
ISBN: 978-1-53614-809-1

Chapter 8

FRESHWATER HARMFUL ALGAL BLOOMS: CAUSES, CHALLENGES, AND POLICY CONSIDERATIONS*

Laura Gatz

SUMMARY

Scientific research indicates that in recent years, the frequency and geographic distribution of harmful algal blooms (HABs) have been increasing nationally and globally. Because the impacts of HABs can be severe and widespread—often with interstate implications—these issues have been a perennial interest for Congress. While algal communities are natural components of healthy aquatic ecosystems, under certain conditions (e.g., increased temperatures and nutrient concentrations), algae may grow excessively, or "bloom," and produce toxins that can harm human health, animals, aquatic ecosystems, and the economy.

* This is an edited, reformatted and augmented version of Congressional Research Service, Publication No. R44871, dated June 20, 2017.

In 2014, a cyanobacterial HAB in Lake Erie affected the drinking water for more than 500,000 people in Toledo, Ohio. In 2016, a massive HAB in Florida's Lake Okeechobee negatively impacted tourism and aquatic life. HABs have been recorded in every state and have become a concern nationwide.

Many types of algae can cause HABs in freshwater systems. The most frequent and severe blooms involve the proliferation of cyanobacteria. Some cyanobacteria species can produce toxins—cyanotoxins—that can cause mild to severe health effects in humans and kill aquatic life and other animals.

HABs can also contribute to deteriorating water quality and ecosystem health. As masses of cyanobacteria or other algae die and decompose, they consume oxygen, sometimes forming "dead zones" where life cannot survive. These areas can kill fish and organisms, such as crabs and clams, and have detrimental economic effects.

Scientists widely consider nutrient enrichment to be a key cause of HAB formation. While nutrients are essential to plants and natural parts of aquatic ecosystems, excessive amounts can overstimulate algal growth. Sources include point sources (e.g., municipal wastewater discharges) and nonpoint sources (e.g., fertilizer runoff from agricultural and urban areas).

Congress, federal agencies, and states have taken steps to address HABs and nutrients that contribute to their occurrence. The Harmful Algal Bloom and Hypoxia Research and Control Act of 1998 (HABHRCA), as amended, established an interagency task force, required the task force to prepare reports and plans addressing marine and freshwater HABs, and authorized funding for research, education, monitoring activities, etc.

In December 2016, the Environmental Protection Agency (EPA) used its authority under the Clean Water Act (CWA) to propose water quality criteria for two algal toxins in waters used for recreational purposes. States use such criteria when developing water quality standards— measures that describe the desired condition or level of protection of a water body and what is needed for protection.

Further, EPA has emphasized the need to reduce nutrient pollution from *all* sources to reduce public health and environmental impacts associated with HABs. The CWA does not authorize EPA to regulate all sources. It authorizes EPA to regulate point (direct) sources of nutrients but does not authorize EPA to regulate nonpoint (diffuse) sources of nutrient pollution.

Some states have developed guidelines for algal toxins, primarily for use in guiding swimming advisories. Also, states have listed waters as impaired, or not meeting water quality standards, for algal blooms or algal toxins. Some of these states have begun to develop Total Maximum Daily Loads (TMDLs)—essentially pollution budgets—to address them. Most states have identified nutrient-related pollution as a priority to be addressed by their TMDLs and/or alternative restoration plans. States rely heavily on financial assistance from EPA in implementing these plans and more broadly, in addressing nonpoint source pollution that leads to degraded water quality and HAB formation. Congress has long provided financial assistance through EPA for regional, state, and local programs through CWA Sections 106 and 319 planning grants, geographic programs (such as the Chesapeake Bay and Great Lakes), and other sources. The President's FY2018 budget request for most of these programs is either eliminated or significantly reduced.

Congress continues to show interest in addressing HABs. This interest has largely focused on funding research to close research gaps identified by scientists and decisionmakers and to coordinate the efforts of federal agencies and their partners to study and address HABs.

INTRODUCTION

Scientific research indicates that in recent years, the frequency and increasing nationally and globally.[1] HABs can be detrimental to human health, animals, aquatic ecosystems, and local economies. In 2014, a major HAB in Lake Erie caused the city of Toledo, Ohio, to issue a "do not drink" order for tap water that left more than 500,000 people without drinking water for two days and had an estimated impact of $65 million in lost benefits.[2] In the summer of 2016, a massive HAB in Lake Okeechobee—Florida's largest freshwater lake—resulted in beach closures, losses to the tourism industry, and negative impacts on marine life. According to the U.S. Environmental Protection Agency (EPA),

---

[1] NOAA, *State of the Science Fact Sheet—Harmful Algal Blooms*, September 2016, http://nrc.noaa.gov/ CouncilProducts/ScienceFactSheets.aspx; U.S. Environmental Protection Agency, "Causes and Prevention," https://www.epa.gov/nutrient-policy-data/causes-and-prevention; National Office for Harmful Algal Blooms at Woods Hole Oceanographic Institution, "Recent Trends: National Changes," July 11, 2016, https://www.whoi.edu/redtide/regions/us/recent-trends. C. B. Lopez et al., *Scientific Assessment of Freshwater Harmful Algal Blooms*, Interagency Working Group on Harmful Algal Blooms, Hypoxia, and Human Health of the Joint Subcommittee on Ocean Science and Technology, 2008, p. 1.

[2] M. Bingham, S. K. Sinha, and F. Lupi, *Economic Benefits of Reducing Harmful Algal Blooms in Lake Erie*, Environmental Consulting and Technology, Inc., submitted to the International Joint Commission, October 2015.

between January 1 and August 12, 2016, states reported at least 266 notices for freshwater HABs, including cautions, warnings, public health advisories, and public health warnings.geographic distribution of harmful algal blooms (HABs) have been

Congress, many federal agencies, states, localities, and other partners have taken and continue to take steps to address the rising trend in HABs and their impacts. However, there are many gaps in current scientific understanding of HABs amongst the research and management communities and considerable debate as to how best to address the issue from a regulatory standpoint. This report explores these issues as they pertain to HABs in freshwater systems. Specifically, it addresses the conditions and activities that contribute to the occurrence of freshwater HABs; steps that Congress, federal agencies—particularly EPA—and their partners are taking to address and mitigate their occurrence; and the current knowledge gaps on this issue. This report is focused on freshwater HABs, not marine or coastal HABs or issues associated with HABs in drinking water supplies.[3]

BACKGROUND

What Are Harmful Algal Blooms?

Algal communities are naturally occurring components of healthy aquatic ecosystems, such as lakes, rivers, and estuaries. However, under certain environmental conditions, such as increased temperatures and nutrient concentrations (e.g., nitrogen and phosphorus), colonies of algae can grow excessively, or "bloom," and produce toxins that pose a threat to human and aquatic ecosystem health and potentially cause economic damage. These HABs sometimes produce discolorations in the water that can appear as scums, paint-like slicks, clotted mats, or foam that may vary

[3] For a discussion of algal toxins in drinking water, see CRS In Focus IF10269, *Algal Toxins in Drinking Water: EPA Health Advisories*, by Mary Tiemann.

in color (i.e., light to dark green, yellow, red, or brown). Even when visible signs of a bloom are absent, however, algal toxins may still cause harmful effects.[4] *Figure 1* shows an aerial view of a HAB that produced visible green scums in Lake Okeechobee, Florida, in July 2016.

Source: USGS, https://www.usgs.gov/news/science-harmful-algae-blooms.

Figure 1. Aerial view of a July 2016 Harmful Algal Bloom in Lake Okeechobee, Florida.

[4] Lopez et al., *Scientific Assessment*, p. 1.

HAB Types and Impacts

While many types of algae can cause HABs in bodies of freshwater, cyanobacteria (sometimes referred to as blue-green algae) typically cause the most frequent and severe blooms.[5] Cyanobacterial HABs pose a threat to human and aquatic ecosystem health and can kill pets, livestock, and wildlife. Some species of cyanobacteria produce toxins, called cyanotoxins, which can cause hepatic (liver-related), neurologic, respiratory, dermatologic, and other symptoms.[6] These may be acute or chronic, mild or severe, and in some cases may be fatal. Humans may be exposed to cyanotoxins by consuming tainted drinking water, fish, or shellfish; swimming or recreating in waters with certain concentrations of cyanotoxins present; or inhaling aerosolized toxins.[7] The cyanotoxins associated with these HABs can contaminate fish, interfere with a variety of recreational activities, and cause other economic and environmental damages.

There are many types of cyanotoxins, which may have multiple variants with a wide range of toxicities.[8] The most commonly occurring and most studied cyanotoxin is microcystin.[9]

Although cyanobacterial HABs are considered to be the most prevalent and toxic types of HABs, blooms of "golden algae" (*Prymnesium parvum*) are an emerging problem and likely the most problematic of non-

---

[5] J. L. Graham, N. M. Dubrovsky, and S. M. Eberts, *Cyanobacterial Harmful Algal Blooms and U.S. Geological Survey Science Capabilities*, U.S. Geological Survey Report 2016-1174, 2016, pp. 1-2.

[6] Centers for Disease Control and Prevention, "Harmful Algal Bloom (HAB)-Associated Illness," https://www.epa.gov/nutrient-policy-data/causes-and-prevention

[7] See footnote 5.

[8] See footnote 5.

[9] See footnote 5 and Keith A. Loftin, Jennifer L. Graham, and Elizabeth D. Hilborn et al., "Cyanotoxins in Inland Lakes of the United States: Occurrence and Potential Recreational Health Risks in the EPA National Lakes Assessment 2007," *Harmful Algae*, vol. 56 (2016), pp. 77-90. Other common cyanotoxins that can have adverse environmental health effects include anatoxins, cylindrospermopsins, and saxitoxins.

cyanobacterial freshwater HAB taxa (i.e., group of related organisms classified as a unit).[10] Golden algal HABs have caused large fish kills worldwide, including millions of fish in Texas. Most of the major fish kills have occurred since 2000. Golden algae thrive in brackish water, such as the rivers and reservoirs found in areas of Texas, Oklahoma, and Wyoming. The toxins produced by golden algae target gill-breathing organisms, such as fish, clams, and mussels. According to information from two states that experience golden algal blooms, there is currently no evidence that golden algal toxins pose a direct threat to humans, other mammals, or birds.[11] Effects of cyanobacterial HABs and golden algae HABs are detailed in *Table 1*.

In addition to the effects of algal toxins on human and animal health, HABs can also contribute to deteriorating water quality and ecosystem health. An over-abundance of cyanobacteria or other algae can block out sunlight and clog fish gills. In addition, as the algae die and decompose, they consume oxygen, leaving waterways in a hypoxic (or low oxygen) state, sometimes forming "dead zones"—areas where life cannot survive due to lack of oxygen. Low oxygen areas can suffocate and kill fish and bottom-dwelling organisms such as crabs and clams. According to EPA, over 166 dead zones have been documented nationwide, including in waterbodies such as the Chesapeake Bay and the Gulf of Mexico.[12] Significant economic losses have occurred as a result of hypoxia.[13]

---

[10] National Office for Harmful Algal Blooms at Woods Hole Oceanographic Institution, "Golden Algae," July 11, 2016, https://www.whoi.edu/redtide/impacts/freshwater/golden-algae

[11] Texas Parks and Wildlife, Texas Commission on Environmental Quality, and U.S. Fish and Wildlife Service, "Texas Golden Alga Facts," http://www.tceq.texas.gov/publications/gi/gi-378.html/at_download/file; Arizona Game and Fish Department, "Golden Alga Frequently Asked Questions," http://www.azgfd.gov/temp/golden_alga_faqs.shtml.

[12] EPA, "Nutrient Pollution, The Effects: Environment," https://www.epa.gov/nutrientpollution/effects-environment.

[13] National Science and Technology Council, Committee on Environment and Natural Resources, *An Assessment of Coastal Hypoxia and Eutrophication in U.S. Waters*, November 2003, pp. 22-23.

Table 1. Selected Freshwater Harmful Algal Bloom (HAB) Taxa and Toxins and Their Effects

HAB Taxa	Toxins	Human Health Effects	Animal Health Effects	Environmental Effects	Economic Effects	Affected Areas in U.S.
Cyanobacteria	Microcystins, cylindrospermopsin, anatoxin-a, saxitoxins	Liver and kidney toxicity, neurotoxicity, paralysis, gastrointestinal effects, dermatitis, respiratory illness	Pet, farm animal, and wildlife mortality; fish kills	Water discoloration, foul odors	Loss of tourism, contamination of drinking water requiring additional expensive water treatment or alternate water sources, taste problems in farmed and wild-caught fish (making them inedible)	Great Lakes and many inland water bodies
Haptophytes (e.g., *Prymnesium parvum,* or "golden algae")	Prymnesins, ichthyotoxins (i.e., fish toxin)	No apparent adverse effects	Kills fish and other gill-breathing organisms, such as clams and mussels	Water discoloration, foam formation	Loss of fishing income, clean-up costs	Alabama, Alaska, Colorado, Florida, Georgia, Nebraska, New Mexico, North Carolina, Oklahoma, Pennsylvania, South Carolina, Texas, and Wyoming

Source: Adapted from National Science and Technology Council, Subcommittee on Ocean Science and Technology, Harmful Algal Blooms and Hypoxia Comprehensive Research Plan and Action Strategy: An Interagency Report, Interagency Working Group on the Harmful Algal Bloom and Hypoxia Research and Control Act, February 11, 2016, p. 49.

Notes: Table includes information only on Freshwater HAB taxa and toxins discussed in this report. For a complete list of freshwater and marine HABs, toxins, and their effects, see the source above, Appendix 1, pp. 49-52.

Factors Contributing to HAB Formation

Many factors may influence the occurrence and prevalence of HABs in freshwater, including nutrient concentrations, water temperature, availability of light, pH, and water circulation. Nutrient enrichment is widely recognized as one of the key causes of HAB formation.[14] Nutrients, such as nitrogen and phosphorus, are essential to plant growth and natural parts of aquatic ecosystems. However, when high levels of nutrients enter a body of water, they stimulate plant and algal growth, which can lead to depletion of dissolved oxygen (as explained above), reduced transparency (i.e., turbidity), changes to the biological community (e.g., loss of sportfish, such as bass), and degradation of the aesthetic appeal of the water (i.e., from odor and scums). This process is called eutrophication.[15]

While some sources of nutrients in water bodies are natural, many anthropogenic activities contribute nutrients to waterbodies from a number of point and non-point sources. Point sources include municipal and industrial wastewater discharges and concentrated animal feeding operations (CAFOs).[16] Nonpoint sources include urban stormwater runoff, failing septic systems, atmospheric deposition of nitrogen from fossil fuel emissions, runoff from fertilized cropland, and manure runoff from cropland, pastures, and animal feeding operations. See *Table 2* for more information on these sources.

Studies also indicate that increased temperatures and changes in frequency and intensity of rainfall associated with climate change may also favor HAB formation.[17] HABs generally proliferate in warmer waters. In

---

[14] See footnote 5 and footnote 4, p. 12.

[15] EPA, "Preventing Eutrophication: Scientific Support for Dual Nutrient Criteria," February 2015, https://www.epa.gov/sites/production/files/documents/nandpfactsheet.pdf.

[16] CAFOs are point sources, as defined by CWA Section 502(14). CAFOs are agricultural operations where animals are kept and raised in confined situations that meet criteria established in EPA's CAFO regulation (40 C.F.R. 122.23). These criteria include specific numbers of confined animals and designation as a significant contributor of pollutants.

[17] H. W. Paerl and J. Huisman, "Climate Change: A Catalyst for Global Expansion of Harmful Cyanobacterial Blooms," *Environmental Microbiology Reports*, vol. 1, no. 1 (February 2009), pp. 27-37. National Science and Technology Council, Subcommittee on Ocean Science and Technology, *Harmful Algal Blooms and Hypoxia Comprehensive Research Plan and Action Strategy: An Interagency Report*, Interagency Working Group on the Harmful Algal Bloom and Hypoxia Research and Control Act, February 11, 2016, p. 12. Jerry M. Melillo, Terese Richmond, and Gary W. Yohe, *Climate Change Impacts in the*

addition, some studies have found that swings between flooding and drought may enhance HAB formation. For example, if intense rainfall is followed by a drought, the nutrients washed into receiving water bodies may remain in them longer, increasing the potential for HABs.[18]

Table 2. Anthropogenic Sources of Nutrients in Water

Source	Source Type	Details
Municipal wastewater discharges	Point	Municipal wastewater treatment plants process wastewater from homes and businesses. The wastewater contains nitrogen and phosphorus from human waste, food, and phosphate-containing soaps and detergents. Some wastewater treatment plants have upgraded their systems and are able to remove more nitrogen and phosphorus than others.
Industrial wastewater discharges	Point	Industrial wastewater treatment plants process wastewater from a variety of manufacturing or industrial activities. According to EPA, certain types of industrial waste tend to possess higher quantities of nutrients, such as those from processors of food, beverages, livestock, and agricultural products.
Concentrated animal feeding operations (CAFOs)	Point[a]	CAFOs are animal feeding operations (see entry below) that meet numeric thresholds for the number of animals they contain and/or meet certain pollution discharge criteria.[b] Manure discharge can contribute nitrogen and phosphorus to waterways.
Animal feeding operations (AFO)	Nonpoint	AFOs are facilities in which livestock or poultry are kept and raised in confinement that meet certain conditions. Manure runoff and wastewater from AFOs can contribute nitrogen and phosphorus to waterways.[c]
Agricultural runoff	Nonpoint	Excess fertilizer applied to crops and fields, animal manure, and soil erosion can all contribute to increased nitrogen and phosphorus entering water bodies during rainfall events.
Urban stormwater runoff	Nonpoint	During rainfall or snowmelt events, water carries nitrogen and phosphorus across paved surfaces

United States: The Third National Climate Assessment, U.S. Global Change Research Program, October 2014, pp. 79, 198, 216.

[18] Paerl and Huisman, "Climate Change," pp. 27-37.

Source	Source Type	Details
		and buildings and into local water bodies or storm drains. Fertilizers, yard and pet waste, and phosphate-containing soaps and detergents can contribute to higher nutrient concentrations in stormwater.
Failing septic systems	Nonpoint	If septic systems are improperly managed, elevated nitrogen and phosphorus levels can be released into local water bodies or groundwater. Common causes of failure include aging, inappropriate design, overloading the system, and poor maintenance.
Fossil fuels	Nonpoint	Combustion of fossil fuels by electric power generation, industry, transportation, and agriculture release nitrogen oxides into the atmosphere. Nitrogen oxides are deposited back onto land and can be washed into nearby waters during rainfall events.

Sources: EPA "Nutrient Pollution: Sources and Solutions," https://www.epa.gov/nutrient pollution/sources-andsolutions; EPA, Office of Water, A Compilation of Cost Data Associated with the Impacts and Control of Nutrient Pollution, EPA 820-F-15-096, May 2015, pp. IV-23.

Notes:

[a] Under CWA Section 502(14), CAFOs are point sources. However, the definition of point sources specifically excludes agricultural stormwater discharges. Therefore, agricultural stormwater discharges from CAFOs are nonpoint sources.

[b] For EPA's regulatory definitions of Large CAFOs, Medium CAFOs, and Small CAFOs, see 40 C.F.R. Section 122.23.

[c] The conditions for an AFO include (1) animals are confined or maintained for a total of 45 days or more in any 12-month period, and (2) crops are not sustained in the normal growing season over any portion of the lot or facility. See 40 C.F.R. Section 122.23.

INCIDENCE AND TRENDS

Scientists largely agree that the frequency and distribution of HABs, the economic losses from them, the types of resources affected, and the number of toxins and toxic species have all increased in recent years.[19] Some scientists note that factors such as better detection methods and increased reporting have contributed to the upward trend. HABs, including cyanobacterial HABs, have been recorded in the waters of all 50 states,

[19] See footnote 1.

with some HABs crossing state lines.[20] *Figure 2* shows the generalized distribution of selected freshwater HAB events (cyanobacterial HABs and golden algal HABs) that took place between 2006 and 2015 across the United States.

The findings of EPA's most recent national assessment of lakes[21] is consistent with other reports of the rising trend in HABs. In EPA's 2012 National Lake Assessment, EPA concluded that there was little change from its 2007 survey of lakes, with two exceptions—trends in algal toxin and nutrient measures.[22] In 2012, EPA and its partners detected microcystin in 39% of lakes, a 9.5% increase from 2007.[23] EPA noted, however, that for both studies, the concentrations of microcystin remained low and rarely exceeded the levels of concern established by the World Health Organization for recreational uses (see "Regulatory Efforts and Guidelines Section"). EPA also found an 8.3% increase in the percentage of lakes in the "most disturbed condition" category when analyzing the density of cyanobacterial cells (i.e., an indicator of risk for exposure to algal toxins because the cells may produce toxins).[24] Finally, EPA found an overall increase in the median concentration of phosphorus across all lakes[25] and a "dramatic" decline (18.2%) in the percentage of lakes with low nutrients and high oxygen levels.[26] These findings are important because in many lakes, phosphorus is considered the limiting nutrient, meaning that the available quantity of this nutrient controls the pace of algal production. It also means that even small increases in phosphorus can lead to very rapid

---

[20] National Office for Harmful Algal Blooms at Woods Hole Oceanographic Institution, "Distribution of HABs in the U.S.," August 15, 2016, https://www.whoi.edu/redtide/regions/us-distribution.

[21] EPA, *National Lakes Assessment 2012: A Collaborative Survey of Lakes in the United States*, EPA 841-R-16-113, December 2016. Every five years, EPA and its partners sample more than 1,000 lakes to inform the agency's National Lakes Assessment, a statistically based assessment of the biological, chemical, physical, and recreational condition of the nation's lakes.

[22] Ibid., pp. 1-2.

[23] Ibid., pp. 1, 18.

[24] Ibid., p. 28.

[25] Ibid., p. 2, 13. The median concentration of phosphorus increased from 20 μg/L in 2007 to 37 μg/L in 2012.

[26] Ibid., pp. 2, 13. In 2012, EPA found 18.2% fewer oligotrophic lakes (i.e., lakes with low nutrients and high oxygen levels) than in 2007.

increases in algal growth.[27] More broadly, the study found that nutrient pollution is a widespread problem across the country. Approximately 35% of lakes have excessive levels of total nitrogen, and 40% of lakes have excessive levels of total phosphorus.[28]

Sources: Graphic adapted from a National Office of Harmful Algal Blooms at Woods Hole Oceanographic Institution map of HABs that occurred between 2006 and 2015 (available at http://whoi.edu/redtide/regions/usdistribution) and an EPA adaptation of the map that reflects input from HAB experts with broad experience in HAB events and reports to the U.S. National Office for Harmful Algal Blooms (see EPA, Draft Human Health Recreational Ambient Water Quality Criteria or Swimming Advisories for Microcystins and Cylindrospermopsin, EPA 822- P-16-002, December 2016, p. 4).

Notes: Each state that has experienced one or more cyanobacterial HABs or golden algal HABs is indicated with a single dot. Larger green ovals mark areas where more widespread cyanobacterial HAB problems occurred.

Figure 2. Generalized Distribution of Selected Freshwater Harmful Algal Blooms (HABs) in the United States.

---

[27] Ibid., p. 12.

[28] The terms total nitrogen and total phosphorus reflect that these measurements include all forms of the nutrient in the sample. For example, total phosphorus includes a measurement of orthophosphate, condensed phosphate, and organic phosphate.

EFFORTS TO ADDRESS HARMFUL ALGAL BLOOMS

Enacted Legislation

Congress has recognized the increasing frequency of HABs and has passed legislation in an effort to address their public health, economic, and environmental consequences. In 1998, Congress passed the Harmful Algal Bloom and Hypoxia Research and Control Act (HABHRCA), which established an Interagency Task Force of Harmful Algal Blooms and Hypoxia.[29] It required the task force to prepare reports assessing HABs and hypoxia with a focus on coastal waters and authorized funding for HAB and hypoxia-related research, education, and monitoring activities.

The Department of Commerce's National Oceanic and Atmospheric Administration (NOAA) chaired the task force. In 2004, Congress reauthorized HABHRCA and expanded it to include assessments of HABs in freshwater.[30] In 2014, Congress again reauthorized HABHRCA and established a national HAB/Hypoxia Program to be maintained by NOAA through the task force.[31] It identified NOAA and EPA as the lead federal agencies for marine and freshwater aspects of the program, respectively, and required additional reports and a comprehensive research plan and action strategy.

In 2015, in response to public safety concerns arising from the Toledo, Ohio, HAB event, Congress passed legislation addressing algal toxins in drinking water. The Drinking Water Protection Act amended the Safe Drinking Water Act to require EPA to develop a strategic plan to assess and manage the risks associated with algal toxins in public drinking water supplies.[32]

The following year, Congress included a provision in the Water Infrastructure Improvements for the Nation (WIIN) Act that required EPA to designate a Harmful Algal Bloom Coordinator to coordinate projects

---

[29] P.L. 105-383.
[30] P.L. 108-456.
[31] P.L. 113-124.
[32] P.L. 114-45.

and activities under the Great Lakes Restoration Initiative involving HABs in the Great Lakes.[33] *Table 3* provides a list and description of the HAB-specific legislation enacted since 1998.

Table 3. Harmful Algal Bloom (HAB)-Related Legislation

Law	Description
Harmful Algal Bloom and Hypoxia Research and Control Act (HABHRCA), 1998 (P.L. 105-383)	• Established an Interagency Task Force of Harmful Algal Blooms and Hypoxia, chaired by the Department of Commerce, to consist of representatives from nine federal agencies, the Office of Science and Technology Policy, the Council on Environmental Quality, and "such other Federal agencies as the President considers appropriate"
	• Required the task force to prepare reports assessing HABs and hypoxia, with a focus on coastal waters • Authorized funding for research, education, and monitoring activities related to HABs and hypoxia: $15,000,000 for FY1999; $18,250,000 for FY2000; and $19,000,000 for FY2001
Harmful Algal Bloom and Hypoxia Amendments Act of 2004 (P.L. 108-456)	• Retained the Interagency Task Force (which the President could have disestablished under the 1998 act) • Expanded the focus of the act to include assessments of HABs in freshwater • Mandated five reports, one of which was to assess current knowledge about HABs in freshwater, including a research plan for coordinating federal efforts to better understand freshwater HABs • Required the task force to complete a scientific assessment of HABs at least once every five years—an assessment that shall examine marine *and freshwater* blooms after the initial of these assessments was complete • Reauthorized funding for research, education, and monitoring activities: $23,500,000 for FY2005; $24,500,000 for FY2006; $25,000,000 for FY2007; and $25,500,000 for FY2008[a]
Harmful Algal Bloom and Hypoxia Research and Control Amendments (HABHRCA) Act of 2014 (P.L. 113-124)	• Established a National HAB/Hypoxia program to be maintained and enhanced by the National Oceanic and Atmospheric Administration (NOAA)[b] through the interagency task force • Identified NOAAb as the lead federal agency with primary responsibility for administering the program and directed EPA to lead the freshwater aspects of the program • Added the Centers for Disease Control and Prevention as a member of the interagency task force

[33] P.L. 114-322.

Table 3. (Continued).

Law	Description
	• Required the interagency task force to develop a comprehensive research plan and action strategy to address marine and freshwater HABs and hypoxia • Required regional reports, including a progress report on Northern Gulf of Mexico hypoxia and an integrated assessment of and plan to address hypoxia and HABs in the Great Lakes Reauthorized funding to implement the program and research plan and action strategy: $20,500,000 for each of FY2014-FY2018[a]
Drinking Water Protection Act (P.L. 114-45)	• Amended the Safe Drinking Water Act to require EPA to develop—and submit Congress—a strategic plan to assess and manage the risks associated with algal toxins in public drinking water supplies[c] • Required EPA to include in the plan steps and schedules for EPA to (1) assess health risks of algal toxins in drinking water, (2) publish a list of toxins likely to pose risks and summarize their health effects, (3) determine whether to issue health advisories for listed toxins, (4) publish guidance on feasible methods to identify and measure the algal toxins water, (5) recommend feasible treatment and source water protection options, and (6) provide technical assistance to states and water systems.
	• Required the Government Accountability Office to report to Congress on federal funds expended for each of FY2010 through FY2014 to examine toxin-producing cyanobacteria and algae or address public health concerns related to harmful algal blooms[d]
Water Infrastructure Improvements for the Nation (WIIN) Act (P.L. 114-322)	• One of the act's provisions directed the EPA Administrator to designate a Harmful Algal Bloom Coordinator to coordinate—with federal partners, Great Lakes states, Indian tribes, and other non-federal stakeholders—projects and activities under the Great Lakes Restoration Initiative involving HABs in the Great Lakes[e]

Source: CRS.

Notes: This table does not include legislation pertaining to cyanobacteria or cyanotoxins in drinking water.

[a] The reauthorization expired in 2008, however, the Consolidated Appropriations Act of 2008 (P.L. 110-161) provided authorizations of $30,000,000 for each of FY2008 through FY2010.

[b] The act specifies that the Under Secretary of the Department of Commerce shall have this role. The Under Secretary of Commerce for Oceans and Atmosphere is the administrator of NOAA.

[c] EPA, Algal Toxin Risk Assessment and Management Strategic Plan for Drinking Water: Strategy Submitted to Congress to Meet the Requirements of P.L. 114-45, 810R04003, November 2015.

[d] U.S. Government Accountability Office, Environmental Protection: Information on Federal Agencies' Expenditures and Coordination Related to Harmful Algae, GAO-17-119, October 2016.

[e] According to EPA, the administrator designated the Great Lakes National Program Office Director as the Harmful Algal Bloom Coordinator.

In addition to HAB-specific legislation, the Clean Water Act (CWA) authorizes EPA to address water quality concerns associated with HABs.[34] The act establishes a system, under Section 303, for states to adopt ambient water quality standards consisting of the designated use or uses of a water body (e.g., recreational, public water supply, or aquatic life) and the water quality criteria that are necessary to protect the use or uses.[35] States then use their water quality standards to determine which waters must be cleaned up, how much effluent may be discharged, and what is needed for protection.

Section 304(a) requires the EPA Administrator to publish and, from time to time, revise water quality criteria that accurately reflect the latest scientific knowledge on the kind and extent of all identifiable effects on human health and the environment that might be expected from the presence of pollutants.[36] These criteria constitute guidance that states use in adopting their water quality standards. As recognized by Section 510 of the CWA,[37] states may develop water quality standards that are more stringent than required by EPA regulations. EPA's water quality standards regulations[38] require that in developing water quality standards, states must adopt water quality criteria that protect the designated use. States are to establish *numerical criteria*—based on (1) EPA's recommended criteria, (2) EPA's criteria modified to reflect site-specific conditions, or (3) other scientifically defensible methods—and establish *narrative* criteria or criteria based on biomonitoring methods where numerical criteria cannot be established or to supplement numerical criteria.

Section 303(d) of the CWA requires states to identify waters that are impaired by pollution, even after application of pollution controls.[39] For those waters, states must establish a Total Maximum Daily Load (TMDL) of pollutants to ensure that water quality standards can be attained. A TMDL is a quantitative assessment of pollution sources and pollutant

---

[34] Clean Water Act, as amended (33 U.S.C. §1251 et seq.)
[35] 33 U.S.C. §1313.
[36] 33 U.S.C. §1314.
[37] 33 U.S.C. §1370.
[38] 40 CFR 131.
[39] 33 U.S.C. §1313.

reductions needed to restore and protect U.S. waters; it is also a planning process for attaining water quality standards. TMDLs may address all pollution sources, including point sources, such as municipal sewage treatment or industrial plant discharges, and nonpoint sources such as urban runoff and agricultural runoff.

Also, Section 118 of the CWA provides that the United States should seek to attain the goals embodied in the Great Lakes Water Quality Agreement of 1978, as amended by the Water Quality Agreement of 1987 and any other agreements and amendments.[40] It tasks EPA to take the lead in the effort to meet the agreement's goals, working with other federal agencies, states, and localities. As seen in the text box, the most recent amendment includes a HAB-related goal.

Great Lakes Water Quality Agreement

The Great Lakes Water Quality Agreement, which was first signed in 1972, is a commitment between the United States and Canada to restore and protect the waters of the Great Lakes.[41] It was amended most recently in 2012 to better identify and manage current environmental issues and prevent emerging environmental issues from affecting the waters of the Great Lakes while upholding and modernizing commitments made in previous agreements. One of the nine objectives of the 2012 agreement is that the waters of the Great Lakes should "be free from nutrients that directly or indirectly enter the water as a result of human activity, in amounts that promote growth of algae and cyanobacteria that interfere with aquatic ecosystem health, or human use of the ecosystem." Although previous agreements had also included efforts to reduce nutrients and prevent excessive algal growth, the 2012 amendments were the first to specifically include cyanobacteria.

FEDERAL AGENCY EFFORTS

Many federal agencies are involved in carrying out various HAB-related activities, including conducting HAB research, monitoring algal

[40] 33 U.S.C. §1269. In 1987 amendments to the CWA (P.L. 100-4) Congress affirmed the goals of the amended Great Lakes Water Quality Agreement of 1978.

[41] Great Lakes Water Quality Agreement, U.S.-Canada, as amended, September, 7, 2012.

toxins and water quality, forecasting HABs, supporting projects to improve water quality, and community outreach efforts. Some in Congress, however, have expressed concern about the activities and expenditures of various agencies and potential redundancies. In the Drinking Water Protection Act (P.L. 114-45), enacted August 7, 2015, Congress directed the Government Accountability Office (GAO) to inventory funds expended by federal agencies to examine toxin-producing cyanobacteria and algae or address public health concerns related to harmful algal blooms. GAO was to recommend ways to improve interagency coordination and reduce duplication of efforts. According to the 2016 GAO report that responded to the mandate, 17 agencies conducted research, monitoring, response, or other HAB-related activities between FY2013 and FY2015.[42] The GAO report provides detailed information on federal agencies' key HAB-related activities, expenditures, and specific statutory authorities, and, thus, this report will not discuss these in detail. Rather, this section identifies the federal agencies involved in a key interagency effort and highlights actions EPA specifically is taking in its role as the leader of freshwater HAB issues.

As previously mentioned, HABHRCA established an interagency task force that is charged with

- promoting a national strategy to help communities understand, predict, control, and mitigate freshwater and marine HAB and hypoxia events;
- enhancing, coordinating, and assessing the activities of existing HABs and hypoxia programs; and
- providing for development of a comprehensive research plan and action strategy.

---

[42] GAO, *Environmental Protection: Information on Federal Agencies' Expenditures and Coordination Related to Harmful Algae*, GAO-17-119, October 2016, p. 5. Note that GAO identified 17 agencies that have conducted research, monitoring, or other HAB-related activities in FY2013-FY2015. However, the report focuses on 12 federal agencies whose data was sufficiently reliable for the purposes of the report.

Table 4 provides a list of the federal agencies and organizations specifically required in HABHRCA to participate on the task force. The reauthorization of HABHRCA in 2014 reconstituted the task force as the Interagency Working Group on the Harmful Algal Bloom and Hypoxia Research and Control Act (IWG-HABHRCA), responsible for maintaining a national HAB/hypoxia program. NOAA and EPA share primary responsibility under HABHRCA for administering the national HAB and hypoxia program, with NOAA leading marine aspects of the program and EPA in charge of freshwater aspects. In addition to agencies listed in the table, the U.S. Army Corps of Engineers is an active member in the IWG-HABHRCA.[43]

Table 4. Interagency Task Force on Harmful Algal Blooms (HABs) and Hypoxia

Representatives Specifically Named in HABHRCA

Agency
Department of Commerce, National Oceanic and Atmospheric Administration (NOAA) [a]
Environmental Protection Agency (EPA)a
Department of Agriculture
Department of the Interior
Department of the Navy
Department of Health and Human Services
National Science Foundation
National Aeronautics and Space Administration
Food and Drug Administration
Office of Science and Technology Policy
Council on Environmental Quality
Centers for Disease Control and Prevention

Source: CRS.

Notes: NOAA and EPA serve as co-chairs of the interagency task force. HABHRCA states that the task force shall also include "such other Federal agencies as the President considers appropriate." According to the interagency task force's 2016 report (see footnote 43), the U.S. Army Corps of Engineers also has a representative on the task force.

[a] NOAA and EPA serve as co-chairs on the task force.

[43] National Science and Technology Council, Subcommittee on Ocean Science and Technology, *Harmful Algal Blooms and Hypoxia Comprehensive Research Plan and Action Strategy: An Interagency Report*, Interagency Working Group on the Harmful Algal Bloom and Hypoxia Research and Control Act, February 11, 2016.

In its role under HABHRCA and the CWA, EPA's efforts to address HABs include coordinating the efforts of multiple entities, developing regulations and guidelines to protect water quality (see "Regulatory Efforts and Guidelines" section), conducting research, providing financial assistance through grants and other agreements, and educating the public.[44] In its coordination role, EPA leads, chairs, or co-chairs several working groups or task forces, including the IWG-HABHRCA, the Inland HAB Discussion Group, the Great Lakes Interagency Task Force, and the Mississippi River/Gulf of Mexico Watershed Nutrient Task Force (Hypoxia Task Force). See *Table 5* for a description of these efforts.

EPA has also conducted internal research on HABs and their toxins focused on water quality (including how different factors such as nutrients, light, temperature, etc., affect HAB occurrence and toxicity), human and ecological health effects, monitoring and analytical methods research, and drinking water treatment research.[45] The agency also provides research grants, such as those provided through the Science to Achieve Results (STAR) program, focused on topic areas including the prediction, prevention, control and mitigation of freshwater HABs and the fate and effects from less-common emerging HABs.[46]

EPA also provides financial assistance to states, tribes, and others to address water pollution, including nonpoint source pollution. Examples of such assistance include nonpoint source implementation grants under CWA Section 319, capitalization grants under the Clean Water State Revolving Fund, and grants under CWA Section 106, which are provided to states, interstate agencies, and tribes to administer programs that prevent, reduce, and eliminate water pollution.

---

[44] Education efforts include communication and outreach, such as newsletters and videos, to increase public awareness regarding the adverse effects of nutrient pollution and HABs.

[45] See footnote 42, p. 35.

[46] EPA "Freshwater Harmful Algal Blooms," Funding Opportunity Announcement Number EPA-G2017-STAR-A1, Office of Research and Development, October 28, 2016.

Table 5. EPA Efforts to Coordinate Freshwater HAB-Related Activities

Coordination Effort	Description
Interagency Working Group on the Harmful Algal Bloom and Hypoxia Research and Control Act (IWG-HABHRCA)	Created after the HABHRCA amendments of 2014, this interagency working group is the primary, government-wide mechanism through which federal agencies coordinate their HAB-related activities and report on specific topics to Congress, such as research plans and action strategies for addressing HABs and hypoxia. The group meets twice a month and is co-chaired by the NOAA and EPA.
Inland HAB Discussion Group	Led by EPA, the U.S. Geological Survey, and the Centers for Disease Control and Prevention to share information among federal, state, local, and industry stakeholders through free webinars, this informal discussion group was created out of an expressed need by federal researchers and state agencies to bridge a communication gap with respect to inland HAB research, monitoring, human and ecological health risk assessment, education, and outreach.
Great Lakes Interagency Task Force	Chaired by EPA, this task force consists of 11 cabinet and other federal agency heads to coordinate the restoration of the Great Lakes. Created by a May 18, 2004, executive order,[a] the task force, among other things, coordinates the development of consistent federal policies, strategies, projects, and priorities pertaining to the restoration and protection of the Great Lakes. According to EPA officials, since 2009, the task force has overseen the implementation of the Great Lakes Restoration Initiative (GLRI),[b] a federally led effort to carry out programs and projects for Great Lakes protection and restoration. In particular, the task force has overseen the development of comprehensive, multi-year action plans that identify goals, objectives, measurable ecological targets, and specific actions for four GLRI focus areas. One of these focus areas is reducing nutrient runoff that contributes to harmful/nuisance algal blooms.
Mississippi River/Gulf of Mexico Watershed Nutrient Task Force (Hypoxia Task Force)	Through this EPA-led task force, federal agencies coordinate with 12 states and a national tribal representative to address hypoxia in the Mississippi River and the northern Gulf of Mexico.

Source: Excerpted from U.S. Government Accountability Office, Environmental Protection: Information on Federal Agencies' Expenditures and Coordination Related to Harmful Algae, GAO-17-119, October 2016, p. 23.

Notes:

[a] Executive Order 13340, "Establishment of Great Lakes Interagency Task Force and Promotion of a Regional Collaboration of National Significance for the Great Lakes," 69 Federal Register 29043-29045, May 18, 2004.

[b] The 114th Congress codified the GLRI in P.L. 114-322, Section 5005. As noted, this provision directed the EPA Administrator to designate a coordinator for GLRI HAB activities.

Regulatory Efforts and Guidelines

EPA and states have also taken steps to address HABs and nutrient loads that contribute to their proliferation through regulatory efforts and guidelines. This section focuses on regulatory efforts and guidelines related to EPA's authorities under the CWA and specifically excludes efforts under the Safe Drinking Water Act.

HABs, Cyanobacteria, and Cyanotoxins

EPA, the World Health Organization (WHO), and many states have developed guidelines for cyanotoxins in recreational waters.[47] These guidelines are summarized in *Table 6* and discussed below.

EPA Guidelines

In December 2016, EPA issued draft recreational water quality criteria for microcystins and cylindrospermopsin for public comment.[48] According to EPA, these criteria reflect the concentrations of two cyanotoxins that would be protective of human health in recreational waters used for swimming or other activities: 4 μg/L for microcystin and 8 μg/L for cylindrospermopsin.[49] EPA suggests that states may consider using the proposed values when determining whether to post swimming advisories in recreational waters and may consider using the same values when adopting new or revised water quality standards.

---

[47] EPA has also developed recommended levels for drinking water for microcystin and cylindrospermopsin through health advisories. See CRS In Focus IF10269, *Algal Toxins in Drinking Water: EPA Health Advisories*, by Mary Tiemann.

[48] As discussed above, CWA Section 304(a) directs EPA to develop and publish and, from time to time, revise criteria for water quality that accurately reflect the latest scientific knowledge. EPA, "Request for Scientific Views: Draft Human Health Recreational Ambient Water Quality Criteria and/or Swimming Advisories for Microcystins and Cylindrospermopsin," 81 *Federal Register* 91929-91931, December 19, 2016. Note this comment period was extended. See EPA, "Extension of Public Comment Period: Draft Human Health Recreational Ambient Water Quality Criteria and/or Swimming Advisories for Microcystins and Cylindrospermopsin," 82 *Federal Register* 10766-10767, February 15, 2017.

[49] EPA, Draft Human Health Recreational Ambient Water Quality Criteria or Swimming Advisories for Microcystins and Cylindrospermopsin, EPA 822-P-16-002, December 2016, p. 1.

Table 6. Summary of EPA, World Health Organization (WHO), and State Guidelines for Cyanotoxins in Recreational Waters

Organization	Microcystin	Cylindrospermopsin	Action
EPA[a]	4 µg/L (DRAFT)	8 µg/L (DRAFT)	Swimming advisory triggered if value is exceeded for one day Recreational waterbody impairment triggered if value is exceeded on more than 10% of days during a recreational season up to one calendar year
WHO[b]	Low < 10 µg/L Moderate = 10-20 µg/L High = 20-2,000 µg/L Very High > 2,000	No guideline	Low: Post on-site risk advisory signs; inform relevant authorities Moderate: Watch for scums or conditions conducive to scums; discourage swimming and further investigate hazard; post on-site risk advisory signs; inform relevant authorities High: Immediate action to control contact with scums; possible prohibition of swimming and other water contact activities; public health follow-up investigation; inform public and relevant authorities
States[c]	0.8 µg/L – 20 µg/L	4 µg/L – 20 µg/L	Variety of actions, including issuing advisories, closing beaches, and increasing sampling

Sources: CRS analysis from the following sources:

EPA, "Draft Human Health Recreational Ambient Water Quality Criteria or Swimming Advisories for Microcystins and Cylindrospermopsin," EPA 822-P-16-002, December 2016; and World Health Organization, "Guidelines for Safe Recreational Water Environments: Volume 1," Coastal and Fresh Waters, 2003.

Notes:

[a] EPA guidelines, as discussed below, are draft criteria and have not been finalized.

[b] Low, moderate, high, and very high refer to the probability of adverse health effects. No additional actions were specifically listed for the "very high" probability category.

[c] Among the 21 states that have quantitative guidelines for cyanotoxins, these values represent the range of the lowest recreational water guideline or action levels that trigger or recommend a health protective action. For additional information on state guidelines and action levels, see table source 1, pp. 11-14 for a list of the lowest recreational water guideline or action level for each state and Appendix B for a more complete list of state guidelines, action levels, and recommended actions.

Many entities—including states, representatives of publicly owned treatment works,[50] agricultural organizations, and environmental groups—provided comments on the draft criteria:[51]

- Some commenters, including states, were supportive of the criteria for purposes of informing swimming advisory decisions but did not support the use of the criteria for developing water quality standards, noting, among other concerns, that cyanotoxins are not a pollutant discharged into waterways but rather result from other pollutants (nutrients) entering waterways and other factors. Environmental groups generally supported EPA's criteria for use in both swimming advisories and development of water quality standards.
- Commenters' opinions varied regarding the proposed concentrations of microcystin and cylindrospermopsin in the draft criteria. Some states felt the levels were appropriate, environmental groups felt they should be more stringent, and other states suggested they are overly protective, particularly when compared to the WHO guideline for microcystin.

Many commenters—particularly states, publicly owned treatment works, and agricultural groups—expressed a number of implementation concerns. One key concern raised was that these criteria, if used for water quality standards, would improperly regulate response organisms rather than a discharged pollutant. Some argued that algal toxins are not a pollutant that CWA permittees discharge. Rather, the discharge of other pollutants, such as excess nutrients, may lead to HAB formation. In its draft criteria document, EPA explained that it does not anticipate states

---

[50] The National Association of Clean Water Agencies, which represents the interests of publicly owned treatment works (i.e., municipal wastewater treatment facilities), provided comments on behalf of its members.

[51] Comments discussed in this section are available in Docket ID No. EPA-HQ-OW-2016-0715, available at https://www.regulations.gov/docket?D=EPA-HQ-OW-2016-0715.

using the criteria alone for permitting purposes, recognizing that cyanobacteria and their toxins are not typically present in permitted discharges. EPA goes on to say:

> Permits are more likely to be written to address point source discharges of the causal pollutants, such as nutrients, on a waterbody-specific or watershed basis, where the permit writer has determined there is a reasonable potential for the causal pollutants in the discharge to cause or contribute to an exceedance of the cyanotoxin standards.[52]

In this regard, some commenters expressed concern that it is not known precisely what level of nutrients will result in a bloom, nor is it understood what factors will trigger the release of toxins. Several commenters suggested that EPA explore these issues further before moving forward with water quality criteria for purposes other than guiding advisory levels for swimming.

Many commenters also expressed implementation concerns regarding monitoring and sampling. According to the Association of Clean Water Administrators[53] many states do not currently have mechanisms in place to adequately sample for the levels of the toxins specified by EPA or lack adequate lab capacity to process increased samples. Some states, publicly owned treatment works, and agricultural groups also commented that the variability of HABs within a body of water and over even short spans of time can make sampling and analysis complicated, particularly when using the data to determine if a water body is impaired. The commenters urged EPA to address these issues in detail before moving forward with the criteria.

WHO Guidelines

In 2003, WHO proposed guideline values for protection from adverse health outcomes associated with cyanobacteria blooms in fresh water used

---

[52] See footnote 49.

[53] The Association of Clean Water Administrators is a nonpartisan, national organization of state, interstate, and territorial water program managers who implement CWA programs.

for recreational purposes.[54] The guidelines are defined at three levels: low, moderate, and high probability of adverse health effects. WHO concluded that a single guideline value was not appropriate because "it is necessary to differentiate between the chiefly irritative symptoms caused by unknown cyanobacterial substances and the potentially more severe hazard of exposure to high concentrations of known cyanotoxins, particularly microcystins." *Table 6* shows the WHO guideline levels for microcystin.[55]

State Guidelines

According to EPA documents, approximately 30 states have implemented cyanobacterial HAB guidelines for recreational waterways as of November 2015. Some of these states use qualitative guidelines only (i.e., visual inspection for blooms rather than quantitative detection methods) or quantitative guidelines for cyanobacterial cell density rather than guidelines for the specific cyanotoxins.

Of the 30 states that have implemented cyanobacterial HAB guidelines, 21 have established numeric guidelines for microcystin or cylindrospermopsin. The levels and associated actions vary considerably among states (see *Table 6*). California has adopted the strictest concentrations for both cyanotoxins (0.8 μg/L for microcystin and 4 μg/L for cylindrospermopsin). Several states have adopted the WHO value of 20 μg/L for microcystin. Only seven states have adopted quantitative guidelines for cylindrospermopsin.

Some states have also added waters affected by algal blooms and algal toxins to their impaired water lists (i.e., Section 303(d) lists) for algal blooms and algal toxins. According to data from EPA's Assessment and Total Maximum Daily Load Tracking and Implementation System, 24 states have listed waters impaired for algal blooms, and two states—California and New Hampshire—have listed waters impaired for algal toxins (see *Table 7*). California has listed only one of its waters as impaired

[54] WHO, Guidelines for Safe Recreational Water Environments: Volume 1, Coastal and Fresh Waters, 2003.

[55] WHO also established guidance values for cyanobacteria and chlorophyll-a.

Table 7. State Algal Bloom and Algal Toxin Impaired Waters Listings and TMDLs

Cause of Impairment	Number of States	Number of Waters Listed as Impaired	Number of Waters with TMDLs
Algal Bloom	24	1,015	52
Algal Toxin	2	67	2

Source: EPA data from the Assessment and Total Maximum Daily Load Tracking and Implementation System as of May 24, 2017.

Note: These data reflect the most recent assessed waters and impaired waters reports provided to EPA by each of the states, ranging from 2004-2014.

for algal toxins and has not yet developed a TMDL. New Hampshire has listed 66 of its waters as impaired for algal toxins and has developed one TMDL that covers two waters.[56] The New Hampshire TMDL establishes a total phosphorus loading target that, if met, is expected to achieve state water quality criteria and thresholds for dissolved oxygen, chlorophyll a (an indicator of algae), and cyanobacteria.[57]

Nutrient Management

Scientists and policymakers widely recognize the need to reduce nutrient inputs to aquatic systems to limit eutrophication and proliferation of HABs. According to EPA, nitrogen and phosphorus pollution is one of the most serious and pervasive water quality problems in the United States.[58] While EPA and states have worked to address nutrient pollution for over a decade, many observers believe more progress is needed to reduce the threat to water quality and public health. EPA has acknowledged that without greater progress, "the successes to date will

[56] According to information provided in New Hampshire's TMDL, lakes were listed as impaired for swimming if surface blooms or "scums" of cyanobacteria were present—even if present only along a downwind shore. AECOM, *Total Maximum Daily Load for Pearly Lake*, Rindge, NH, prepared for EPA Region 1, August 2014, p. 2-5.

[57] Ibid, p. 1-1.

[58] EPA, Office of Water, *FY2016-2017 National Water Program Guidance*, EPA 420-R-15-008, April 2015, p. 15.

likely be outpaced by the rapidly increasing population and the resulting increase in the rate and impact of nitrogen and phosphorus pollution."[59]

According to EPA, 45 states identified nutrient-related pollution as a priority to be addressed by TMDLs and/or alternative restoration plans in setting long-term priorities for their CWA Section 303(d) programs.[60] To date, more than 8,600 nutrient-related TMDLs have been established, primarily by states, to guide nutrient reduction efforts in more than 5,800 waterbodies.[61]

In 2016, EPA issued a memorandum with a renewed call to states and stakeholders to intensify their efforts, in collaboration with EPA, to reduce nutrient pollution.[62] The memorandum emphasized EPA's support for state planning or implementation of watershed-based, multi-stakeholder projects to reduce the impacts to public health from nitrogen and phosphorus pollution contributing to HABs. EPA listed and described key elements of its plans for working with partners and stakeholders over the next several years, including prioritizing watersheds and setting load reduction goals, developing numeric nutrient criteria, reducing point sources of nutrient pollution, reducing nutrient loads from nonpoint sources, and providing financial and technical assistance.

For almost two decades, EPA has expressed support for developing numeric criteria for nutrients. In a memorandum issued in 2011, EPA stated that "it has long been EPA's position that numeric criteria targeted at different categories of water bodies and informed by scientific understanding of the relationship between nutrient loadings and water quality impairment are ultimately necessary for effective state

---

[59] EPA, Office of Water, Actions to Help States Address Barriers to Numeric Nutrient Criteria Implementation, EPA 820-F-13-011, August 2013.

[60] Joel Beauvais, Deputy Assistant Administrator, EPA, memorandum to State Environmental Commissioners, State Water Directors, "Renewed Call to Action to Reduce Nutrient Pollution and Support for Incremental Actions to Protect Water Quality and Public Health," September 22, 2016, https://www.epa.gov/nutrient-policy-data/working-partnershipstates-address-phosphorus-and-nitrogen-pollution-through.

[61] Ibid., p. 4. EPA has approved approximately 70,000 TMDLs in total since 1995 to address impairments from many different pollutant types. See https://ofmpub.epa.gov/waters10/attains_nation_cy.control#tmdls_by_state.

[62] See footnote 60.

programs."[63] To this end, EPA has provided 30 states with technical assistance for numeric nutrient criteria development through its Nutrient Scientific Technical Exchange Partnership and Support Program.[64] To date, 23 states have adopted numeric criteria into their water quality standards for nitrogen and/or phosphorus for at least one of their water bodies.[65] In 2013, EPA outlined barriers to numeric nutrient criteria implementation and actions to help states address them.[66] The barriers included, among other things, an inability to reduce nonpoint source loads of nitrogen and phosphorus and problems implementing water-quality based limits.

EPA has also emphasized the need to focus on reducing nutrients from all sources—both point and nonpoint sources.[67] Under the CWA, EPA has authority to regulate discharges from point sources.[68] However, the CWA does not authorize EPA to regulate nonpoint sources. EPA can influence activities of nonpoint sources only through use of grants and funding—such as CWA Section 319, which addresses nonpoint source pollution through state-run nonpoint pollution management programs—and related technical assistance.[69] Through such programs, states may, for example, ask farmers or ranchers to use alternative methods in their operations to prevent fertilizers from reaching streams and may provide funds to help them install on-farm pollution management systems or practices. In its document addressing barriers to numeric nutrient criteria, EPA proposed actions to address them, including continuing to collaborate with the U.S.

---

[63] Nancy K. Stoner, Acting Assistant Administrator, EPA, memorandum to Regional Administrators, Regions 1-10, "Working in Partnership with States to Address Phosphorus and Nitrogen Pollution through Use of a Framework for State Nutrient Reductions," March 16, 2011, https://www.epa.gov/nutrient-policy-data/working-partnership-statesaddress-phosphorus-and-nitrogen-pollution-through.

[64] See footnote 60.

[65] EPA "State Progress toward Developing Numeric Nutrient Water Quality Criteria for Nitrogen and Phosphorus," https://www.epa.gov/nutrient-policy-data/state-progress-toward-developing-numeric-nutrient-water-quality-criteria. Only one state (Hawaii) and three territories have adopted numeric criteria for nitrogen and phosphorus for all applicable water types (lakes/reservoirs, rivers/streams, estuaries).

[66] See footnote 59.

[67] See footnote 60.

[68] CWA §402; 33 U.S.C. §1342.

[69] While the 319 program is voluntary at the federal level, states may include regulatory components in their 319 programs.

Department of Agriculture to leverage resources for conservation practices[70] and to better quantify the environmental results of best management practices and other efforts, continuing to implement the Section 319 grant program, and addressing the challenges of manure management by working with large animal growers and poultry integrators to develop sustainability agreements and practices that reduce nutrient pollution.[71]

Some observers argue that the voluntary nature of controlling nonpoint sources is a key challenge in developing and implementing TMDLs, a primary tool that states are employing to address nutrient pollution. Farming and forestry groups have long been concerned about how their activities might be addressed in TMDLs and whether they might be subject to CWA regulation of some sort, even though the act does not provide EPA with regulatory authority over nonpoint sources. Municipalities and industries contend that regulating only point sources imposes disproportionate requirements on their operations, especially in waters that are impaired both by point and nonpoint sources.

Federal Financial Assistance

Recognizing that a critical role for EPA in addressing nutrient pollution is supporting watershed-based efforts at the state and local level, in its 2016 memorandum, the agency stated that the Office of Water would continue to provide financial assistance to states through CWA Section 106 and Section 319 grant programs and the Clean Water State Revolving Fund (CWSRF) Program, as well as Section 604(b) planning grants,[72] Wetland Program Development grants, and grants targeted toward specific

---

[70] A number of U.S. Department of Agriculture agencies provide support through education, outreach, and research, while federal funds are provided through conservation programs to help agricultural producers adopt best management practices for nutrient reduction. Examples of such programs include the Environmental Quality Incentives Program and the Conservation Stewardship Program. For more information see CRS Report R43919, *Nutrients in Agricultural Production: A Water Quality Overview*, by Megan Stubbs.

[71] See footnote 59.

[72] CWA Section 604(b) requires states to reserve a small portion of each its CWSRF allotment each fiscal year—1% for most states—to carry out planning under CWA Sections 205(j) and 303(e). States generally use Section 604(b) grants to fund regional comprehensive water quality management planning activities to improve local water quality.

geographic locations, such as the Chesapeake Bay, Great Lakes, and other water bodies. The President's FY2018 budget request for EPA proposes that funding for these programs, with the exception of the CWSRF, be eliminated or significantly reduced (see *Table 8*).[73] The proportion of funds provided to nonpoint source pollution projects through the CWSRF program is relatively minor compared to the amount provided to publicly owned treatment works for infrastructure projects. As reported by EPA, 96% of the cumulative assistance provided through the CWSRF as of 2016 has been provided to publicly owned treatment works; only 4% was provided to nonpoint source pollution projects and National Estuary Program projects.[74] It is unclear how, considering EPA's long-standing emphasis on using these programs to address nonpoint source pollution, the FY2018 budget would support the goals of the agency in its efforts to reduce nutrient pollution, and ultimately reduce the occurrence and frequency of HABs.

Table 8 presents a comparison of the President's FY2018 budget request with the FY2016 and FY2017 enacted appropriations for selected grants and programs referenced above that include funding support for addressing nonpoint source pollution. These grants and programs are funded within the EPA State and Tribal Assistance Grants and the Environmental Programs and Management appropriations accounts.

RESEARCH GAPS

In addition to the challenges of reducing nutrient pollution contributing to HABs, scientists widely recognize key research gaps that hinder the ability to prevent, predict, minimize, and suppress HABs. In reauthorizing HABHRCA in 2014, as discussed above, Congress directed NOAA—

[73] EPA, Fiscal Year 2018 Justification of Appropriation Estimates for the Committee on Appropriations, EPA-190-K17-002, May 2017, pp. 114, 456-457.

[74] EPA, 2016 Annual Report – Clean Water State Revolving Fund Programs, 2016, https://www.epa.gov/sites/production/files/2017-03/documents/2016_cwsrf_annual_report.pdf.

Table 8. FY2016 and FY2017 Enacted Appropriations and FY2018 Requested for Selected EPA Grants and Programs That Include Assistance to Address Nonpoint Source Pollution

(Dollars in Millions)

EPA Appropriations Acct. and Grants/Programs	FY2016 Enacted P.L. 114-113	FY2017 Enacted P.L. 115-31[a]	FY2018 Budget Request
State and Tribal Assistance Grants Acct.			
Infrastructure Assistance			
Clean Water State Revolving Fund Program	$1,393.9	$1,393.9	$1,393.9
Categorical Grants			
CWA Section 106 Grants	$230.8	$230.8	$161.3
CWA Section 319 Grants	$164.9	$170.9	$0.0
Wetland Program Development Grants	$14.7	$14.7	$10.2
Environmental Programs & Management Acct.			
Geographic Programs[b]			
Chesapeake Bay Program	$73.0	$73.0	$0.0
Great Lakes Restoration Program	$300.0	$300.0	$0.0
Gulf of Mexico Program	$4.5	$8.5	$0.0
Lake Champlain	$4.4	$4.4	$0.0
Long Island Sound	$3.9	$8.0	$0.0
Puget Sound Program	$28.0	$28.0	$0.0
South Florida Program	$1.7	$1.7	$0.0
San Francisco Bay	$4.8	$4.8	$0.0
Lake Pontchartrain	$0.9	$0.9	$0.0
Southern New England Estuaries	$5.0	$5.0	$0.0
Other Geographic Activities	$1.4	$1.4	$0.0

Sources: FY2016 and FY2017 enacted amounts are as reported in the "Explanatory Statement" accompanying the Consolidated Appropriations Act, 2017 (P.L. 115-31), as published in the Congressional Record, vol. 163, no. 76-Book II (May 3, 2017), https://www.gpo.gov/fdsys/pkg/CREC-2017-05-03/pdf/CREC-2017-05-03-bk2.pdf. Under Division G, see discussion under "Title II—Environmental Protection Agency," beginning p. H3883, and in the funding table beginning p. H3920. FY2018 requested amounts are as reported in EPA's Fiscal Year 2018 Justification of Appropriations Estimates for the Committee on Appropriations, EPA-190-K-17-002, May 2017, pp. 114, 456-457, https://www.epa.gov/planandbudget/fy-2018-justification-appropriation-estimates-committeeappropriations.

Notes: Most of these grants provide financial assistance that supports and addresses many water quality improvement projects and other activities in addition to nonpoint source pollution.

[a] FY2017 amounts do not reflect rescissions.

Although some of the geographic programs' water bodies are estuarine in nature (outside the scope of this report), their watersheds include freshwater sources. Funding may therefore support efforts to reduce nonpoint source pollution contributions to the freshwater sources.

through the IWG-HABHRCA—to prepare a comprehensive research plan and action strategy to address marine and freshwater harmful algal blooms and hypoxia. The February 2016 task force report[75] includes a discussion of the key challenges in HAB and hypoxia management and discusses the many gaps in the research and management communities' knowledge of HAB and hypoxia events. Federal agencies—including USGS,[76] EPA,[77] and NOAA[78]—and research efforts sponsored by these agencies cite similar gaps and areas needing continuing research. The following text box summarizes some of these key research areas.

HAB Research Gaps

- Ability to predict the timing, species composition, and toxicity of HABs (including the environmental triggers for toxicity)
- More information on the influence that excess nutrients and other factors (e.g., climate) play in the occurrence and distribution of HABs
- Need to strengthen and integrate new and existing monitoring programs (i.e., expand river, stream, and watershed monitoring of nutrients and toxins, coupled with increased modeling information)
- Need to develop standardized and validated detection and analysis methods for algal toxins
- Need for more effective methods for suppressing, mitigating, or controlling HABs
- Need to understand less-common HAB species that produce toxins
- Better understanding of the human health effects of HAB exposure
- Better understanding of the environmental, economic, and social impacts of HABs
- Improvements in public outreach and communication strategies and tools

---

[75] See footnote 43.

[76] See footnote 5.

[77] Lesley V. D'Anglada, EPA, "Editorial on the Special Issue 'Harmful Algal Blooms (HABs) and Public Health: Progress and Current Challenges,'" *Toxins*, vol. 7 (October 2015), pp. 4437-4441. Also, EPA's solicitation for grant proposals regarding HABs lists specific research gaps the applications should address. EPA "Freshwater Harmful Algal Blooms," Funding Opportunity Announcement Number EPA-G2017-STAR-A1, Office of Research and Development, October 28, 2016.

[78] NOAA, *State of the Science Fact Sheet—Harmful Algal Blooms*, September 2016, http://nrc.noaa.gov/CouncilProducts/ScienceFactSheets.aspx.

LEGISLATION IN THE 115TH CONGRESS

Congressional interest in HABs has largely focused on funding further research and coordinating the efforts of federal agencies and their partners to study and address HABs. In the 115th Congress, S. 1057 would reauthorize HABHRCA for FY 2019-2023, add the Army Corps of Engineers to the interagency task force, and allow the administrators of NOAA (marine) or EPA (freshwater) to declare a HAB or hypoxia event as an event of "national significance." Such a determination would prompt authority for the administrators to provide funding to the affected state or local government.

S. 129—reported by the Senate Committee on Commerce, Science, and Transportation on March 30, 2017—would reauthorize the National Sea Grant Program and make funds available for HAB research through that program. The bill would authorize grants for university research on several targeted topics, including "the biology, prevention, and forecasting of harmful algal blooms." Several bills in the 114th Congress would have done the same (H.R. 1900, H.R. 4394, and S. 3282).

The Great Lakes and Fresh Water Algal Bloom Information Act (H.R. 1893) in the 115th Congress would require NOAA to create an electronic database of research and information on the causes of, and corrective actions being taken with regard to, algal blooms in the Great Lakes, their tributaries, and other surface fresh waters and for other purposes. A similar bill was previously introduced in the 114th Congress (H.R. 349).

An additional bill (H.R. 2137) seeks to address concerns arising from the HAB outbreak that occurred in Lake Okeechobee, Florida, during the summer of 2016. Record rainfalls and an impending hurricane season prompted the U.S. Army Corps of Engineers to release water from the lake into rivers and estuaries in an attempt to keep the lake from overflowing. The releases occurred at a time when a lake-wide HAB was forming, allowing nutrients and algae to move downstream,[79] eventually reaching area beaches

---

[79] Florida Sea Grant, "Florida Sea Grant Funds Research on Massive Algae Bloom on Lake Okeechobee," July 21, 2016, https://www.flseagrant.org/news/2016/07/florida-sea-grant-responds-massive-algae-bloom-lake-okeechobee/

and requiring beach closures and cleanups that had substantial impacts on the local and state economy. Florida's governor requested a federal state of emergency declaration, which the Federal Emergency Management Agency (FEMA) denied because FEMA determined that supplemental federal assistance under the Robert T. Stafford Disaster Relief and Emergency Assistance Act[80] was not considered appropriate for the event.[81] H.R. 2137 would direct the President to treat a harmful algal bloom "caused by certain activities of the Federal Government" as an emergency for the purposes of the Stafford Act.

CONCLUSION

Recent HAB events highlight the public health, economic, and environmental consequences that communities in the United States may continue to experience, perhaps on a more frequent basis. EPA, NOAA, and other federal agencies are working together to conduct important HAB-related research in an effort to close the gaps in the scientific and management community's understanding of how best to prevent, predict, minimize, and suppress HABs. EPA, states, and their partners are working to identify and restore waterbodies that are affected by HABs and the excess nutrients that contribute to their formation. Congress has passed legislation to help drive and fund research efforts and improve collaboration among the many federal agencies involved in HAB-related activities. Moving forward, Congress may be interested in oversight and implementation of the Administration's efforts to implement HABHRCA and other HAB-related authorities. While Congress, federal agencies, and states are taking steps to address HABs, many observers assert that further action is needed to make progress that outpaces the growing consequences of nutrient pollution.

---

[80] 42 U.S.C. §5121 et seq.

[81] Letter from W. Craig Fugate, Administrator, Federal Emergency Management Agency, U.S. Department of Homeland Security, to Rick Scott, Governor of Florida, July 15, 2016.

Most observers agree that further research is needed to understand the most appropriate way to predict, minimize, and suppress HAB outbreaks, including whether and how to regulate algaltoxins. These advocates assert that Congress should ensure that adequate funding is available for such research. To control HABs, some advocate for regulation of nonpoint source pollution, arguing that point sources are disproportionately regulated and that nonpoint sources are the larger contributors to nutrient pollution. Instead of regulation, some argue that EPA and other federal agencies should continue to focus on collaborative, voluntary watershed-level efforts to address nonpoint source pollution that contributes to HAB formation and that Congress should ensure that financial assistance for these efforts continues.

Controlling nonpoint sources of excess nutrients that contribute to HAB formation is challenging. They are diffuse and pervasive and often attributable to many sources and activities rather than a single cause. Yet, scientists generally agree that the current trends in over-enriched waters and HAB events cannot be corrected without addressing nonpoint source nutrient pollution in a significant way and that controlling point sources alone is not enough. Given the consequences of HABs and the difficulty in controlling nonpoint sources of the nutrients that contribute to their formation, challenges and issues associated with HABs are likely to remain of interest to Congress.

INDEX

B

C

D

E

F

G

H

I

J

K

L

T

Environmental Change and Sustainability: Management, Challenges and Research Opportunities

Editor: Lydia Henderson

Series: Environmental Research Advances

Book Description: *Environmental Change and Sustainability: Management, Challenges and Research Opportunities* examines a variety of underlying conceptions regarding sustainability, providing a number of different views on its causes and effects.

Softcover ISBN: 978-1-53613-376-9
Retail Price: $82

Groundwater Contamination: Performance, Limitations and Impacts

Editor: Anna L. Powell

Series: Water Resource Planning, Development and Management

Book Description: Groundwater contamination is nearly always the result of human activity. This book focuses on the performance, limitations and impacts of groundwater contamination.

Softcover ISBN: 978-1-53611-003-6
Retail Price: $82

Water Resources: Systems, Management and Investigations

Editor: Rachel A. Lambert

Series: Water Resource Planning, Development and Management

Book Description: This book provides new research on water resources. Chapter One discusses water resource management and intelligent monitoring systems for reservoirs. Chapter Two describes actions applied to the management of water resources in the region of La Plata in Argentina.

Softcover ISBN: 978-1-53610-976-4
Retail Price: $95

Improving Oil Spill Prevention and Response, Restoring Jobs, and Ensuring Our Energy Security: Recommendations from the National Commission on the BP Deepwater Horizon Oil Spill and Offshore Drilling

Editor: Clay A. Lindeman

Series: Environmental Remediation Technologies, Regulations and Safety

Book Description: This book was reprinted from a public document and records from the US Government. It is a collection of hearings that took place in Washington DC, by the Committee on Transportation and Infrastructure, House of Representatives on February 11th 2011.

Softcover ISBN: 978-1-53613-634-0
Retail Price: $95